PAH Publishing
Classic Muscle Car
By the Numbers Restoration Guide Series

Pontiac Firebird Restoration Guide 1967-1969

Joe Moore

First Published in 2005 by PAH Publishing International
711 Hillcrest, Monett, Missouri 65708 USA

The information in this book is true and complete to the best of our knowledge. All recommendations are made without any guarantee on part of the author or Publisher, who also disclaim any liability incurred in connection with the use of this data or specific details. Always use jack stands and exercise caution when working on your automobile.

We recognize that some words, model names and designations, for example, mentioned herein are the property of the trademark holder. We use them for identification purposes only. This is not an official GM publication.

All drawings and illustrations in this book are the property of PAH Publishing unless othewise stated.

Firebird Restoration Guide 1967-1969
p. cm.
Includes index
ISBN:978-0-9747730-3-2
1. Pontiac restoration 2. Firebird Restoration. 1 Title

Cover Photo: 1967 Firebird coupe owned by Gray Hoagof Pittsburg KS at the Roaring River State Park in Cassvile, MO

Printed and bound in the United States of America

Contents

Acknowledgments

This book would have not been possible without the assistance of the following people. David Osborne of Jim Osborne Reproductions who, generously supplied the decals that were used in this guide. They offer a great selection of quality decals for your Firebird. Benny Henderson, a good friend that allowed us to go take photos of the salvaged Firebirds in his yard. Another good friend Ed Witte of CST Auto Salvage, And a special thank you to; Michael Craig, Glen and Cheryl Meyer, James and Linda Martin, Jackie and David Green, Matt and Mindi French and Barry Weber, all of whom attended the POCI Nationals, where we got the opportunity to enjoy viewing and photographing their cars for use in this guide. Also thanks to the Pontiac Oakland club who hosted the event. This book is dedicated to Mom, who has always been there for us. There are those that are born to love you and those that choose to love you. I am glad that she chose to love me.

Preface

Its funny but as a Firebird enthusiast, you should never laugh or sneer at a Mustang, nor should you think any less of the GTO. In an era that the Beatles were changing to the way we grooved to music, and the way we wore our hair. These two models forever changed the way that GM brass looked at things. Believe it or not there was time when speed, and sporty looks were not viewed as sellable. In fact it was just the opposite, oh there were the few specialty models like the 409 from Chevrolet, the Max wedges from Mopar , Ford's 427 and of course Pontiac's highway brawler the 421 S.D, but these were rare models sold as only handfuls and viewed by the corporate brass as those for "extremists" who wanted to go racing. The Mustang placed the primary goal on sporty looks. Aimed at a growing youth market-the baby boomers, while GTO went the other way and placed the objective of performance as the number one point, which was aimed at the same market. In 1967 GM would put these two together and create the first true pony car, sporty looks and packing a powerful punch. And there was no car better at it then the Pontiac Firebird. Packing up to 400-ci, and with a suspension that could out corner the Mustang. Pontiac set the stage, and invited all on comers. They offered something for everyone, form the mild timid there was the Firebird six, for those wanting a touch of the European flair there was the Sport 4-bbl six-cylinder Sprint, to those wanting full fisted punch in the form of the 400 Ram Air. And for those wanting it all there was the Trans Am, which was introduced in 1969. Just like a sleek muscle car pasting you in the night the days of they came. The Firebird would continue on till it was laid to rest in 2002. Each generation of Firebird would have it's own personality, and the first generation was that of brute strength. This guide is meant to review that strength and like the bird that it is name after raise again to become the warrior of the street.

Information for this guide was drawn from Pontiac factory material inclined parts catalog and service bulletins. Part numbers listed may not match the ones on your car, as part numbers differed for the various vendors that made the part, also when parts ran out other parts were substituted. Every effort was made to ensure that information used in this guide is correct. Therefore, no responsibility can be taken on the action of the authors or the publisher if any inaccuracies exist.

Chapter 1

Documentation and Decoding

Vehicle Identification Number

Regardless of the sub model all first generation Pontiac Firebirds came with a Vehicle Identification Number. This number was used to register when it was sold but was also used to track the car during warranty repair. It is crucial that the VIN on your car matches that on your title, or you cannot properly register your car. Even a Trans Am convertible is not worth anything if the paperwork is messed up. While on second and later generations of Firebirds sub-models, like Trans Am, can be determined by the VIN this is not true for pre 1970 models. The only model that can be determined for the 1967-1969 model years is Firebird, and body style.

Location for the VIN depends on the model year. For 1967 the VIN is reversed stamped into a stainless tag that is riveted to the hinge pillar of the left door. For 1968 and 1969 models the tag is found attached to the driver's side instrument panel. While the location may change the way it decodes is the same all three years.

Division line will always be the number "2" which indicates the Pontiac line in the General Motors family. For the most part the car line will be the code"23" which decodes as Firebird. The exception to the rule is the 1967 models. Some model have used the code "24" for V-8 models, and "25" and "26" for Custom models with a six cylinder and V-8 powerplants respectively. The reason for this is because of different style numbers that were used on the Fisher body tag.

Only two body styles were available: "37" for two-door hardtop and "67" for convertible. Model year will be the last digit of the model year, for example 7-1967, 8-1968 9-1969. Now to put to rest an old street tale. Early first generation Firebirds were not built along with Chevrolet Camaros. Camaros were built at the Van Nuys plant in California, and the Norwood assembly plant in Ohio. The 1967 and 1968 Firebird was built at only one location the Lordstown in Ohio, which was coded with the letter "U" in the VIN. Beginning in 1969 (however, a few late 1968's may have been built here as it was switched over), Firebirds were also built at the Van Nuys assembly plant in Los Angeles, CA and will have the letter code "L" in the VIN plate. No first generation Firebirds were built at the Norwood assembly plant.

Most serial numbers began at 100001 at each assembly plant each year. However, there were exceptions like the six cylinders, which began at 600001. So remember that the code 600101 means that it was the 101st six cylinders built, not the 600101st. The V-8 model would be code as 100101 for the 101st car built.

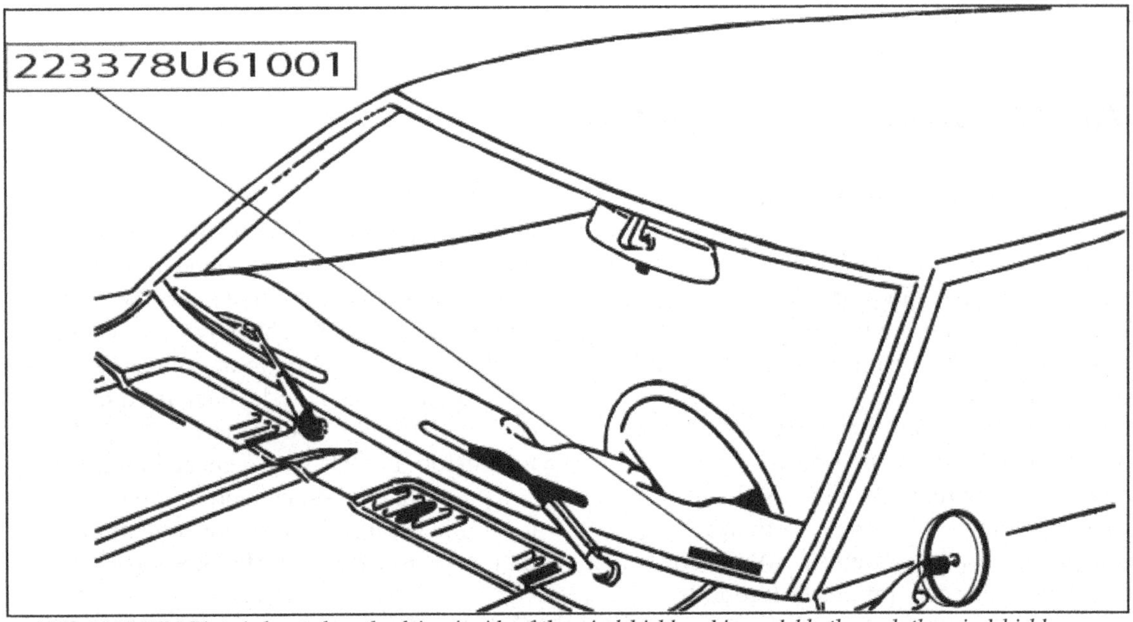

1968-1969 VIN Plate is located on the driver's side of the windshield and is readable through the windshield

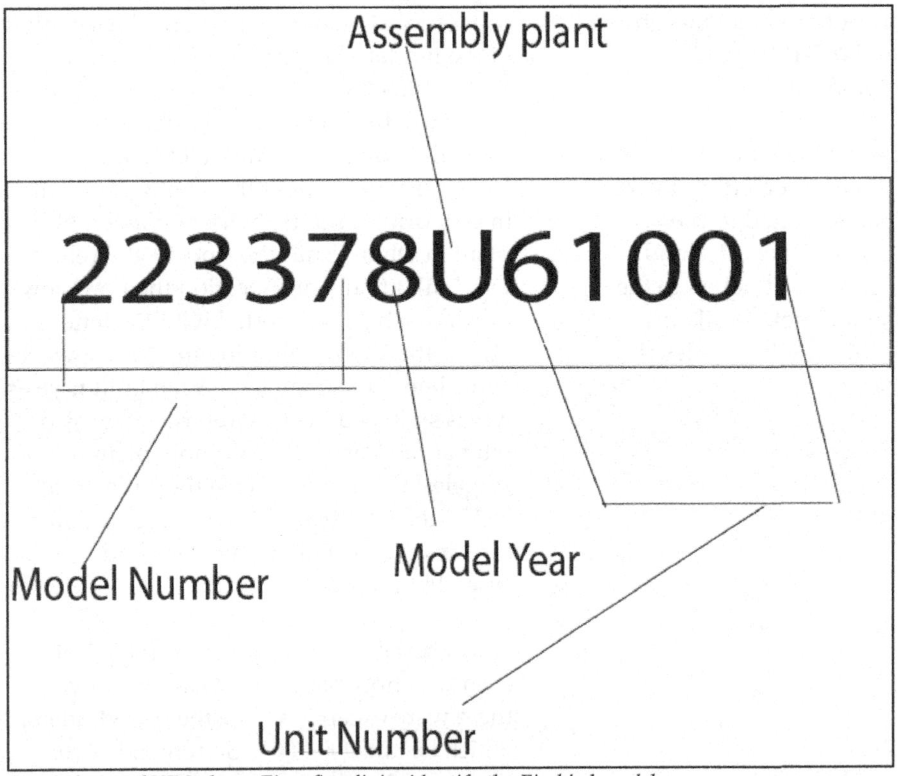

Break down of VIN plate. First five digits identify the Firebird model

1967 Body Tag Style Numbers

Body Style	Standard	Custom
Hardtop	2337*	2537*
	2437	2637
Convertible	2367*	2537*
	2467	2637

1967 Accessories Group Codes

Code	Option	Code	Option
Group 1		Group 3	
D	Power Top	B	Rear Window Defogger
E	Tinted Glass (all)	D	Aux. Lighting Group
L	Fold Down Rear Seat	H	Sprint Option
O	Two-Tone Paint	F	Firebird 400
T	Power Bucket Seat	J	Custom Trim
W	Tinted Windshield	K	Courtesy Lamp Group
X	Power Windows	M	326 H.O.
Y	Power Bench Seat	S	Interior Décor
Group 2		Group 4	
B	3-spd Floor Shift	F	Outside Remote Mirror
E	Air Conditioning	Q	Full-size Spare
G	Center Console	Group 5	
H	Delete Heater	B	Appearance Group
K	Dealer installed A/C	C	Shoulder Belts
L	4-spd	J	Under Hood Light
M	2-spd automatic	M	Rear Seat Center Belt
R	Rear seat speaker	Y	Deluxe Seat Belts
S	Rear Antenna	Z	Custom Shoulder Belts.
T	Power Antenna	Note there are cases where the option code for example C60 fro air conditioning was used instead of these codes. These codes had no meaning except to the factory.	
U	8-track player		
V	Verba-Phonic Rear Seat Speaker		
W	3-spd Automatic		

Body Number Plate

Also known as a data plate or trim plate it was installed by the Fisher Body Group before final assembly of your car, and contains a wealth of information about your car especially the 1967 tag.

Information locked up in code for all model years includes the beginning build date, model style, and exterior color and interior trim color and style. Certain option that appears, that were installed on your 1967 Firebird may also appear in code on the bottom line. However, you should be aware of the fact that discrepancies have occurred on the plate. The number one reason for these disagreements is that the assembly plant ran out of supplies, and instead of getting custom trim you got standard trim, even though the trim tag reads it received custom trim. Another reason, although it is less common, and usually occured with the exterior color, is the customer changed their mind. Instead of Cameo Ivory they now want Solar Red, even though the tag says it was originally painted white, the original color is red. Do not fret, most tags are correct.

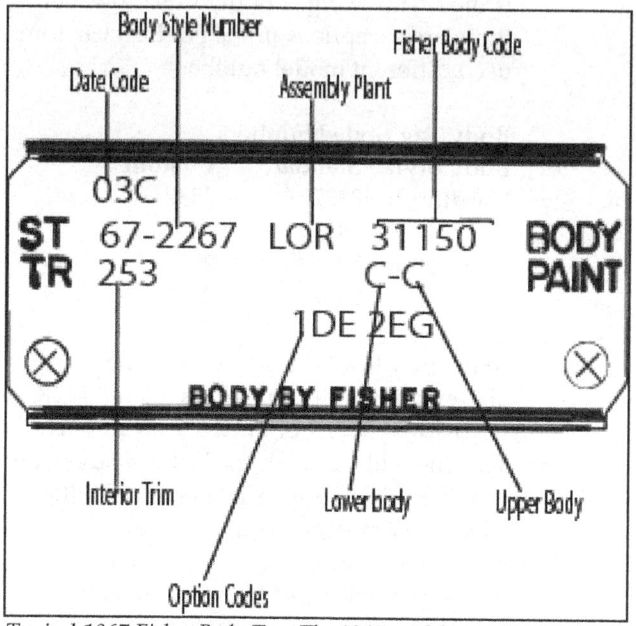

Typical 1967 Fisher Body Tag. The 1967 models are the only Firebirds to have the option codes appear on the tag.

1967 Interior Trim

Color	Bench Seat	Bucket Seat
Blue	270	250
	265	255
Gold	271	251
	267	257
Red	N/A	252
		258
Black	272	253
	269	259
Turquoise	N/A	256
Parchment	260	254

The build date is located on the top line of information on the 1967-early 1969 Body tags and the bottom line on late 1969 models. Although its location changed the way it decodes did not. The first two digits of the code is the month 01-January to 12 December, this is followed by a letter code that indicates the number of week in the month (A-1st through E-5th). The code 04C would translate the third week of the fourth month or in other words the fourth week of April.

The letters ST stand for Style Number. It is follow by a body code that begins with the last two digits of the model year (67-1967,68-1968 and 69-1969) this is followed by a dash or space and the model number, usually this is the first five digits of the VIN, however there are exceptions in the Firebird Custom used different model numbers.

Body Tag Style Numbers

Body Style	Standard	Custom
Hardtop	2337*	2537*
	2437	2637
Convertible	2367*	2537*
	2467	2667

In the middle of this line will be the abbreviation for the assembly plant "LOR" Lordstown or "L" or "LOS" for Van Nuys. The line will end with the Fisher Body code number. This number will not match the VIN, the assembly plant used it and is the sequential number built at that plant. So the numbers 140002 and 140003 were built together. While the numbers in the VIN do not always indicate that the cars were built together but instead were ordered in line.

1967 Exterior Color Codes

Code	Color Name
A	Starlight Black
C	Cameo White
D	Montreux Blue
E	Fathom Blue
F	Tyrol Blue
G	Signet Gold
H	Linden Green
K	Gulf Turquoise
L	Mariner Turquoise
M	Plum Mist
N	Burgundy
P	Silver Glaze
R	Regimental Red
S	Champagne
T	Montego Cream
1	White Fabric Top or Vinyl roof
2	Black Fabric top or vinyl roof
7	Cream vinyl top

The next line will begin with the letters "TR" for Trim. A three-digit code that indicates the color and style of interior trim will appear here. At the end of this line is the code for the exterior color of the car. You will see a double code; most cars will have the same code letter (or numbers on 1969 models) repeated. This means that the upper and lower exterior colors are the same. However, a two-tone paint job, or more likely a car with a colored vinyl or convertible top will have different codes. The first code is the lower part of the car, while the second is the roof color.

Accessories codes will appear on the bottom line of the 1967 Body Number Plate. These codes are broken into several different groups. Each group is identified with a number that will begin the group, which is followed by the code letters. For example: the code 2EGL on the tag would indicate that air conditioning, center console and a 4-speed were installed on this car. Now before fore warned that not all tags used these codes and some were misstamped.

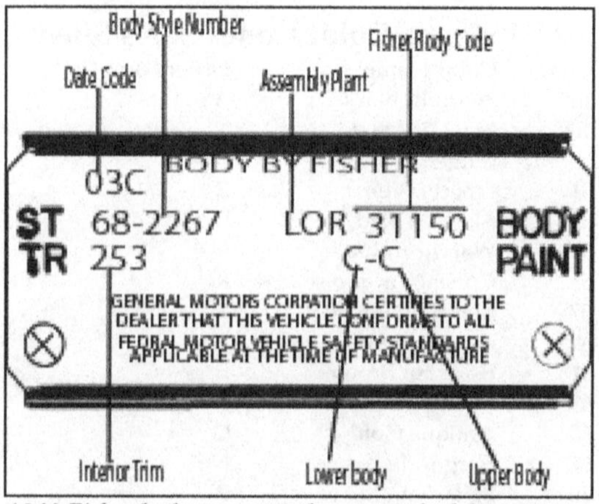

1968 Fisher body tag, note the removal of the option codes that had to make way for the federal mandated wording

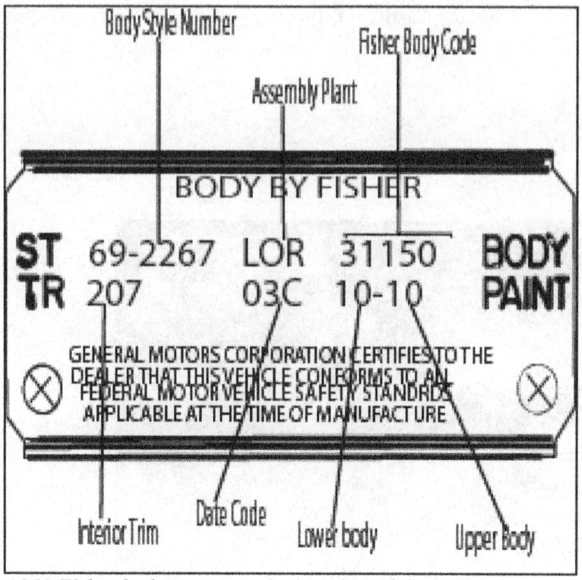

1969 Fisher body tag; note the location of the build date. Some early models may use the 1968 style and date location.

1968 Interior Trim

Color	Standard		Custom	
Type of Seat	**Bench**	**Bucket**	**Bench**	**Bucket**
Teal	N/A	250	N/A	255
Gold	N/A	251	N/A	257
Red	N/A	252	N/A	258
Black	272	253	268	259
Turquoise	N/A	261	N/A	256
Parchment	273	262	275	260

1968 Exterior Color Codes

Code	Color Name
A	Starlight Black
C	Cameo White
D	Alpine Blue
E	Agenda Blue
F	Nordic Blue
G	April Gold
K	Meridian Turquoise
L	Aleutian Blue
N	Flambeau Burgundy
P	Silvermist Green
R	Solar Red
T	Primavera Beige
V	Nightshade Green
Y	Mayfair Maize
1	White Fabric Top or Vinyl roof
2	Black Fabric top or vinyl roof
5	Teal convertible top
8	Gold Convertible top

1969 Interior Trim

Color	Standard		Custom	
Type of Seat	**Bench**	**Bucket**	**Bench**	**Bucket**
Blue	N/A	200	N/A	210
Gold	N/A	202	N/A	212
Green	N/A	206	N/A	216
Red	N/A	N/A	N/A	214
Parchment	227	207	N/A	217
Black	228	208	N/A	218
Gold	Leather			293

1969 Exterior Color Codes

Code	Color Name
10	Starlight Black
40	Mayfair Maize
50	Cameo White
51	Liberty Blue
52	Matador Red
53	Warwick Blue
55	Crystal Turquoise
57	Midnight Green
59	Limelight Green
61`	Espresso Brown
63	Champagne
65	Antique Gold
67	Burgundy
69	Palladium Silver
72	Carousel Red
73	Verdoro Green
76	Goldenrod Yellow
86	Claret Red
87	Windward Blue
88	Nocturne Blue
89	Castilian Bronze
2	Black top
3	Dark Blue top
5	Parchment top
8	Dark Fawn top
9	Dark Green top

1969 Exterior Color Codes Sales Sheet

Code	Color Name	Order Form Code
10	Starlight Black	A
40	Mayfair Maize	Y
50	Cameo White	C
51	Liberty Blue	E
52	Matador Red	R
53	Warwick Blue	D
55	Crystal Turquoise	K
57	Midnight Green	M
59	Limelight Green	H
61`	Espresso Brown	B
63	Champagne	S
65	Antique Gold	G
67	Burgundy	N
69	Palladium Silver	P
72	Carousel Red	T
73	Verdoro Green	Q
76	Goldenrod Yellow	W
86	Claret Red	L
87	Windward Blue	F
88	Nocturne Blue	V
89	Castilian Bronze	J

1969 Trans am.

Chapter 2

Frames

All 1967-1969 Firebirds are built on GM's F-body format, which it shares with its GM cousin the Chevrolet Camaro. This platform consists of unitized construction. A sub frame supported the front-end sheet metal, front suspension, engine, and in some cases the transmission. Because of this design it is essential that the sub-frame be properly aligned, to ensured correct suspension location. If you are buying or restoring a Firebird you believe has been in a collision it is a must that you check the alignment, and if needed have it realigned. This is a repair that is best left to professionals, as the over all strength of the frame is essential to the way the car will handle and body panels will align. If the frame is beyond repair then it is best to replace the whole unit. Be aware that two different frames were used and a unit from a 1968 will not properly align up on a 1969 model, or vise versa.

Front Sub-Frames Part Numbers

1967	3912560
1968	9793077
1969	546703

The transmission assembly was supported at the rear of the sub frame by a cross-member. The part number for this part is dependent on the transmission type. Both the frame assembly and the transmission cross-member should be painted first with a good quality primer. Then both assemblies should be painted with semi-gloss black (Chassis Black) for detail. The original unit was not painted but was unfinished black, but this quickly faded. To give your show car a more show appearance and protect the under side use the black paint.

Transmission Cross-Member Part Numbers

3-speed Manual	3792133		
4-speed Manual	3792133		
2-speed Automatic	9792574	1967-69	Col. Shift
	9798953* 9798982**	1967-69	Console
3-speed Automatic	479795	1967-69	

*-Frame ** Transmission

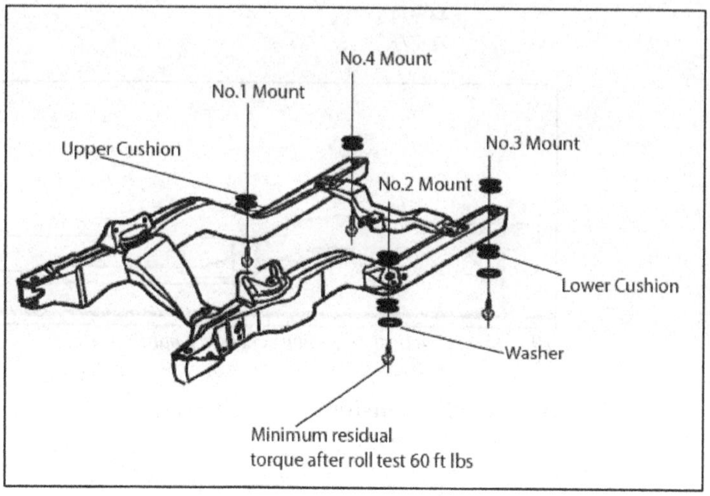

The front subframe attached to the body at 4-points the two front locations are shown here.

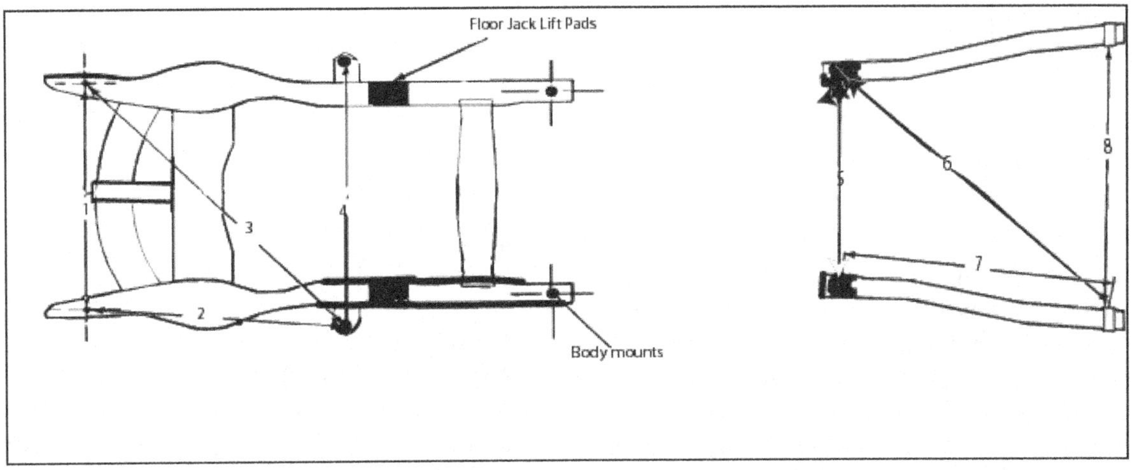

1967 frame horizontal alignment measurements. Dimensions to gauge holes are measured to dead center of the holes and flush to adjacent surface metal unless otherwise specified

Horizontal Dimensions

Fig Ref.	Dimension	Fig Ref.	Dimension
1	38-3/16	5	33-1/2
2	34-15/16	7	55-3/16
3	54-1/8	7	55-3/16
4	44-9/16	8	42-7/8
5	75-7/8		

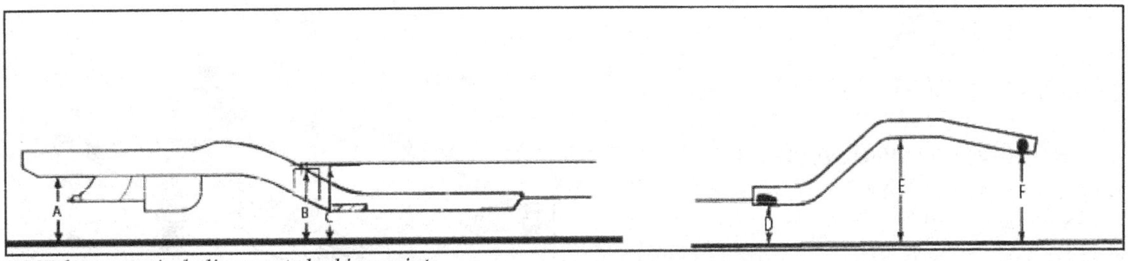

1967 frame vertical alignment checking points.

Vertical Dimensions

A	11-15/16	B	13.00
C	13-13/16	D	6-5/16
E	11-3/4	F	15-11/16

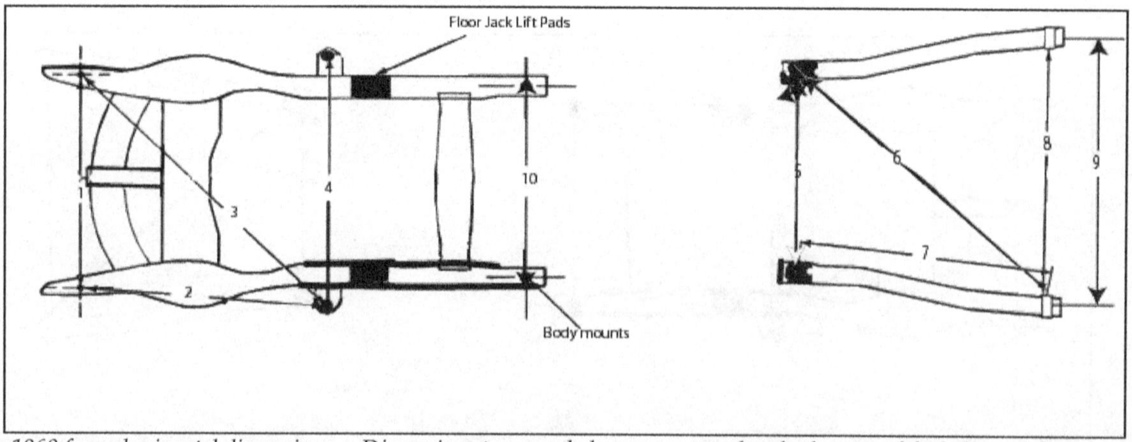

1968 frame horizontal dimensions. . Dimensions to gauge holes are measured to dead center of the holes and flush to adjacent surface metal unless otherwise specified

Horizontal Dimensions

Fig Ref.	Dimension
1	38-1/4
2	35-1/4
3	54-3/16
4	44-9/16
5	33-1/2
6	66-11/16
7	55-3/16
8	42-7/8
9	44-7/8
10	33-3/4

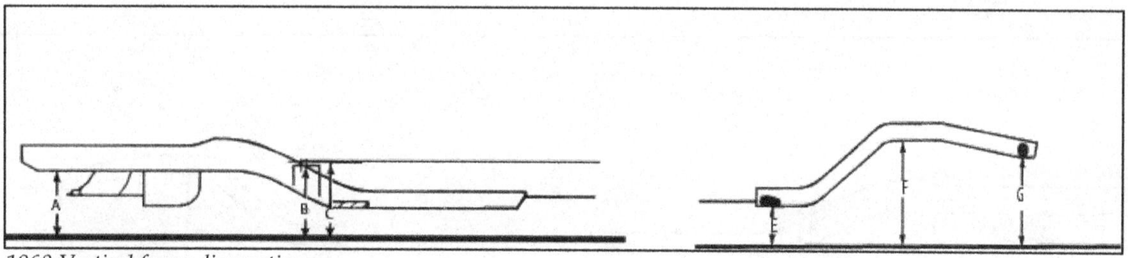

1968 Vertical frame diamentions.

Vertical Dimensions

A	11-15/16	B	13.00
C	13-13/16	D	9-1/8
E	6-15/16	F	15-11/16
G	11-3/4		

1967-1968 Body Dimensions

Length Overall	188.8	Wheelbase	108.1
Width Overall	72.6	Track-Front	59.0
Height	52.0	Track Rear	60.0

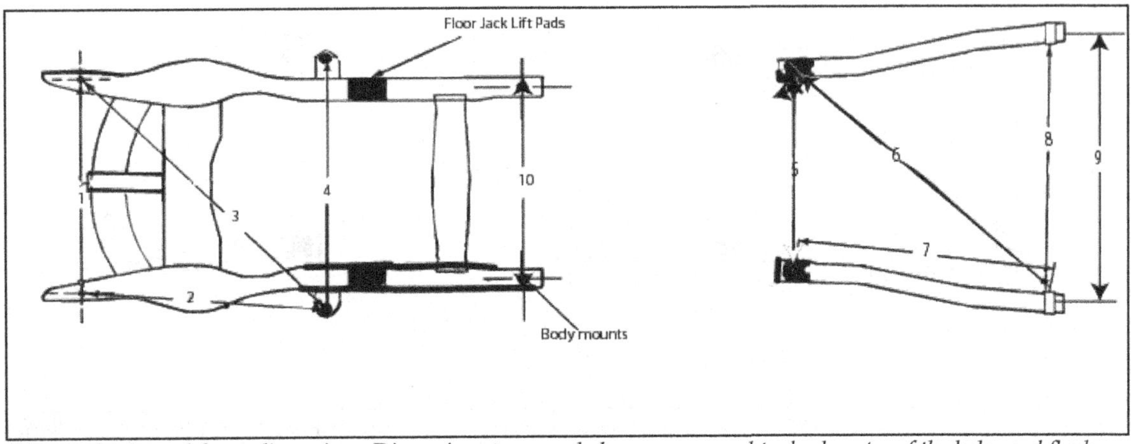

1969 Frame horizontal frame dimensions. Dimensions to gauge holes are measured to dead center of the holes and flush to adjacent surface metal unless otherwise specified

Horizontal Dimensions

Fig Ref.	Dimension	Fig Ref.	Dimension
1	38-3/16	7	66-11/16
2	34-15/16	8	42-7/8
3	54-1/8	9	44-7/8
4	44-9/16		
5	55-3/16		
6	42-7/8		

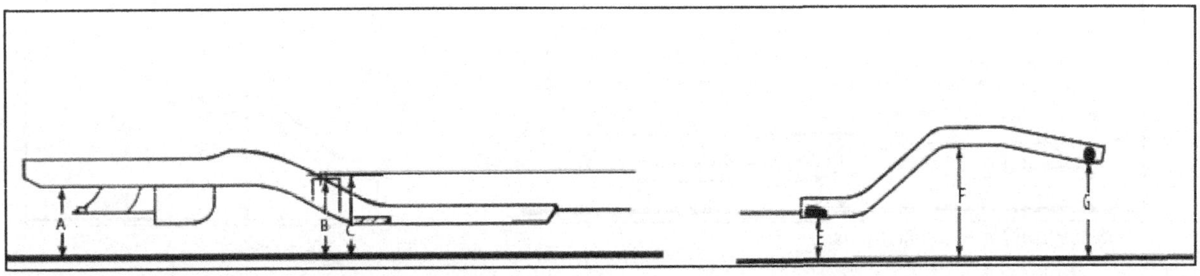

1969 Frame horizontal frame dimensions. Dimensions to gauge holes are measured to dead center of the holes and flush to adjacent surface metal unless otherwise specified

Vertical Dimensions

A	11-15/16	B	13.00
C	13-13/16	D	6-15/16
E	11-3/4	F	15-11/16

Chapter 3

Suspension and Steering Systems

Front Suspension

All Firebird's both the first and the second generation models used the same type of front suspension system. That consisted of wishbone type upper and lower control arms with a coil spring wedged in between. Base of proven designs like the GTO's, it is highly stable and except under great stress provided a good handling ride. Different rates of springs were used according to the suspension options and powerplant that was ordered. For example, springs were higher rated in a car with a 400—ci V-8 than those with a 350-ci V-8.

The same pair of control arms were used all three years on all models. The upper arms were listed as part numbers 9796890- right-hand and 9796891-left-hand and the lower control arms as part numbers 9794994-right-hand and 9794995-left-hand. Originally the parts were dipped, at the part manufacturing plant, in a semi-gloss black paint to prevent the parts from rusting on the shelves. Dipping these parts in paint should not be an option to the restorer, as the results will be less than expected. To duplicate the like new appearance, spray the arms with chassis black. For the upper arm the entire assembly should be painted black. The lower control all of the arms (except approximately 2-inches at the ball joint end) should be painted chassis black. Originally this end was the end that was held when the part was dipped and was left natural and unpainted, painting that end detail gray can duplicate this look. For other details use a small brush and dab silver model paint on the heads of the rivets that hold the ball joint. The mounting bolts had a flat black finish.

SPRINGS

Several different rates of springs were used according to the options that were installed on the car. Springs should have a natural appearance, which can be duplicated by painting them detail gray. There are cases where I have heard of owners keeping the part natural then painting it with a clear coat of paint. The problem with this method is that the clear paint has tendency to give the parts an improper gloss.

SHOCK ABSORBERS, FRONT

Only two different types of shocks were used throughout the run of early Firebirds, all of which were made by Delco. The body of the shock assembly should have a semi-gloss dove gray appearance. The color of this shade is unique, but several different reproduction companies offer the correct shade, one such company is Eastwoods's.

Front Shock Absorber Part NUmbers

Part Number	Year	Usage
3192955	1967	Except Firm Ride
3192953	1967	Firm Ride
3192977	1968	Except Firm Ride
3182037	1968	Firm Ride
5353143	1969	Except Firm Ride
5353144	1969	Firm Ride

SWAY BARS

Various sizes were used according to suspension package and/or model. The bar should have a semi-gloss black appearance; duplicate this by painting it chassis black. However, some bars were left natural, this can be duplicate by painting it detail gray.

Front Sway Bar Part Numbers

Part Number	**Year**	**Diameter/Usage**
3908373	1967-68	11/16-in.
3955778	1969	13/16-in.
3958466	1969	Trans Am

Torque Specifications Front Suspension						
Upper Control Arm		**Lower Control Arm**			**Shock Absorber**	
Mounting Nut	Lub fitting	Mounting Nut	Lub fitting	Rubber Bumper Nut	Mounting nut	Lower screw
45-55 lbs ft.	40-60 lbs in.	70-90 lbs ft.	40-60 lbs in.	11-140 lbs in.	60-120-lbs. in.	15-25-lbs ft.
Sway Bar						
Mounting Nut	Bracket Screws					
15-20 lbs ft.	10-15 lbs. ft.					

Rear Suspension

All first generation Firebirds utilized a simple leaf spring style suspension. This type of suspension is economical. The drawback to this type of suspension is that it is more prone to axle wind up and wheel hop, if it is not properly set up. This was especially true of the 1967 models due to the single leaf design. This was changed with the 1968 models when multiple leaf springs and a staggered shock arrangement were used. The Chevrolet Camaro used the same arrangement. Now while the 1968-1976 Nova and its clones the Ventura II and Buick Apollo used a similar arrangement, and the springs will correctly bolt up your Firebird, these springs have a slightly different arch and will not provide the correct ride height.

The springs should have a natural appearance. This can be duplicated by painting them detail gray, which gives the illusion of unpainted steel. The springs were rubber mounted at the axle and frame attaching points to reduce road harshness and were portioned by locating pins butt-welded to the spring leaf. The spring shackles and brackets were painted semi-gloss black (chassis black) the u-bolt that holds the springs together should have a natural appearance.

A special 20-in. long radius rod was used only on certain 1967 models, sort of a traction bar. This bar helped somewhat with wheel hop. The bar and it's bushings are serviced as a whole unit. It is held in place to brackets with two nuts and bolt at the front and four bolts and nuts at the rear bracket. Remove the front bracket and radius rod together, then remove the front and rear bushing bolts and nuts on the workbench to remove the bracket. The radius rod part number 9789801 should have a semi-gloss black finish for detail, as should the brackets. When installing the rod be sure you tighten the front bracket bolts on the car and with the weight of the car on the wheels, in other word the wheels on the ground and not on a jack stand.

Number of Radius Rods Used

Transmission	6-Cyld 1-bbl	6-Cyld. 4-bbl	V-8
Manual	1	2	2
Automatic	None*	1	1

*-1- if with 3.23 rear axle

1967 REAR SPRING PART NUMBERS								
	STANDARD						FRIM RIDE	
Model Body Style	6-Cyld. 1-bbl	6-Cyld 4-bbl	326-ci 2-bbl	326-ci 4-bbl	400-ci Ram Air 400 H.O.	400-ci Sport Option	6-cyld.	V-8
Hardtop	3925815	5925816	5925816	5925816	3925817	3925817	3925817	3925817
Convertible	3909942	3925817	3925817	3909941	3909940	3909940	3909940	3909940

1968 REAR SPRING PART NUMBERS								
	STANDARD						FRIM RIDE	
Model Body Style	6-Cyld. 1-bbl	6-Cyld 4-bbl	350-ci 2-bbl	350-ci 4-bbl	400-ci Ram Air 400 H.O.	400-ci Sport Option	6-cyld.	V-8
Hardtop	9793458	482548	482548	9797303	9797303	9797303	9797303	546482
Convertible	9793459	9789635	9789635	9793461	9789637	9789635	9789637	546482

1969 REAR SPRING PART NUMBERS								
	STANDARD						FRIM RIDE	
Model Body Style	6-Cyld. 1-bbl	6-Cyld 4-bbl	350-ci 2-bbl	350-ci 4-bbl	400-ci Ram Air 400 H.O.	400-ci Sport Option	6-cyld.	V-8
Hardtop	482548	482548	482548	9789636	9789636	482548	9789636	9789636
Convertible	9789635	9789635	9789635	9789637	9789637	9789635	9789637	9789637

Number of Leafs in Springs

Part Number	**No.**	**Notes**
482548	6	
9789635	6	
9789636	4	
9789637	4	
9793458	5	
9793459	6	
9797303	5	

SHOCK ABSORBERS, REAR

Only two different types of shocks were used throughout the run of early Firebirds. The type of suspension regulated the shock usage; those with the firm ride package used higher rated shocks than those with the standard type suspension. Delco made both types. Note that some models like the H.O. models came standard with the firm ride package and it would be incorrect for theses models to use the standard style shocks. Handling would also greatly suffer. All were made by Delco and the body of the shock assembly should have a semi-gloss dove gray appearance, which will match the front shocks.

Rear Shock Absorber Part Numbers

Part Number	Year	Usage
3192957	1967	Except Firm Ride
3192976	1967	Firm Ride
3192085	1968	Except Firm Ride
3192086	1968	Firm Ride
3182146	1969	Except Firm Ride
3182147	1969	Firm Ride

Rear Suspension Torque Specifications

Application	Torque	Application	Torque
Nut, Shock to lower mount	60 ft lbs	Screw, Shock Absorber bracket to floor pan	8-ft-lbs.
Nut, Upper shock absorber mount to bracket	8-ft-lbs.	Nut, Spring and shock absorber anchor plate attaching	40-ft. lbs
Nut, Spring front eye bolt	75-ft. lbs	Screw Front Bracket to Frame	25-ft. lbs
Nut, Rear Spring Shackle	50-ft. lbs.		

Steering Systems

Firebird steering systems were based on the same unit that was used on the Tempest line, and they have similar aspects. The only difference is the amount of travel. The Firebird has a 35-degree of movement from center position of the pitman shaft in either direction. Based on recalculating-ball steering gear box, which was susceptible to wear. Rebuilding this should be must in a restoration of your first generation Firebird, to restore the originally handling of the car.

Basically two different styles were offered in the Firebird line the standard manual steering gearbox and the power-assisted unit. There were several different units available, based on the power plant and the model, but these are the two basic designs.

Steering Gear Boxes

Part ID #	Usage	Model Years
7802532	P/S 35-degree	1967-1968
7805654	VRS (except T/A)	1968-1969
7807056	Trans Am	1969
7801340	6-cyld or VB-8 without P{/S or a/c	1967-1969
7806396	V-8 without P/S with A/C	1967-1969
9786985	1967V-8	With A/C 1/2x53-1/2 in.
9789939	1967-69 V-8	With A.I.R. and A/C 12x53 in.
9786984	1968-69 6-cyld.	With A/C 1/2x51-1/2 in.

To remove the steering gearbox, you will have to raise up the car and support it with jack stands. Next remove the pitman arm from the pitman shaft, which may require a special tool. Next disconnect the steering shaft flange from the flexible coupling. Then remove the three mounting bolts that hold the gear box to the front sub frame, note this will require a rust dissolving solution like Liquid Wrench, and may even require other means such as heat or cutting the bolts off. Use cutting bolts only as a last resort as this will require that you find the original type of bolts, and finding usable original type bolts is a task that is hard to do.

If the steering gearbox was okay, meaning that the car handle they way it was supposed to, then you can get back with carefully and thoroughly cleaning the case in solvents to remove the decades of crud that have formed, steel wool and fine wire brush may be required. Be extra careful in cleaning the cover, as this unit is made of aluminum and is more easily damaged. You can then prep and refinish the gearing box.

The Saginaw steering gear unit had a special blue gray finish. This finish can be very hard to duplicate. The only company we know that reproduces the correct shade is OEM paints. The cover, which should have a dull aluminum finish, can usually be done with a careful cleaning or painting it semi gloss aluminum The built date code appears on the cover. The cover holding bolts should have a natural unpainted finish, which can be duplicated by painting them detail gray or were sometime flat black, which can be accomplished with a blacking kit. The end plug should also have a dull silver appearance. The end plug retainer ring should have a natural finished steel appearance.

If the gearbox was not in good working order, it will require a rebuild. This involves disassembling the unit and rebuilding the insides, the rebuild kits are fairly straightforward and usually contain instructions on how to rebuild the unit. Because different rebuild kits may have different instructions, we will not list the details of that here.

Several different steering columns were used according to the type of transmission and steering that was ordered in the car. The columns have a unique pattern of finish. The area under the hood, and visible is usually painted semi-flat black. The pitman arms, idler arms, intermediate rods, and inner and outer tie-rod ends were left natural from the factory; painting them cast iron gray can duplicate this. When painting tie rods be careful not to get paint on the rubber seals, as paint can deteriorate them, use masking tape to seal this area off when painting. The tie rod clamps should have a natural or flat black finish both styles were used. Tie rod nuts should be painted with silver model paint to emulate the fresh factory appearance. The holding carter pins should have a dull silver appearance.

The 1967-1969 Firebird steering linkage part numbers are as follows. Intermediate rod part number 3953221- 1967-68 and 5697552- 1969 6-cyld and 7806827 1969 V-8. Pitman arm are 3953225- 1967 Standard steering- 6cyld; 3908379 1967 6-cyld power steering; 9789156 1967 V-8 standard steering; 9789157 1967 V-8 power steering 3953217 1968 6 cyld standard steering; 3953231 1968 6-cyld. Power steering, except V.R.S.; 3908379 6-cyld with V.R.S.; 9791823 1968 V-8 with standard steering; 9791824 1968 V-8 with power steering except V.R.S.; 9789157 1968 V-8 with V.R.S.; 7806626 1969 with standard steering and 7806617 1969 with power steering. The idler arms listed were different each year 1967-3908383; 1968-3917579; 1969 7806613 (with support and bushings). Tie rod clamps were held with 3/8x 1-5/8 in hex head bolts and nuts, they should be torqued to 140-ft-lbs.

A total of three different power steering pumps were used on the 1967-1969 Firebird models. The part number depends on certain options like emission controls and engine sizes. Another option that will effect pump usage is the 1968-69 Variable steering option which can power assist variable steering ratio from 16:1 to 12:1, and reduced the number of turns of the steering wheel from 3- ½ with regular power steering to just 3 full turns, note this measurement is with the pitman arm disconnected. The pump assembly was painted 60-degree gloss black. The mounting brackets that supported the pumps should also be painted 30-degree gloss black

Power Steering Pumps

Part Number	Usage	Notes
5698070	1967 6-cyld	W/O A.I.R.
	1967 V-8	On 6-cyld.
	1968-69 all	Except V.R.S
7801239	1967 6-cyld.	With A.I.R.
7805161	1968 all	With V.R.S.
	1969 all	

Power Steering Pump Pulley Part Numbers

Part Number	Usage	Notes
9778555	1967-69 6-Cyld	W/O A/C or A.I.R.
9778562	1967-69 6-Cyld.	With A/C or with A.I.R
		2-groove pulley
9788972	1967 6-cyld	With A.I.R. and A/C

Power Steering Pump Belt Part Numbers

Part Number	Usage	Notes
9789021	1967 6-cyld	With A.I.R. 3/8x31-1/4 in.
9779516	1967 6-cyld	With A/C 3/8x27-5/8 in.
9789230	1967-69 6-cyld	W/O A.I.R. or A/C 1/2x52 in. Double wrap
9786985	1967V-8	With A/C 1/2x53-1/2 in.
9789939	1967-69 V-8	With A.I.R. and A/C 12x53 in.
9786984	1968-69 6-cyld.	With A/C 1/2x51-1/2 in.

Steering Column Part Numbers

Part Number	Transmission Type	Shifter Location	Mode Year	Notes
7805856	Three-speed auto.	Column	1967	Without tilt wheel
7805847	Automatic	Column	1967-68	With tilt wheel
7805848	All	Floor	1967-68	With tilt wheel
7805856	Automatic	Column	1967-68	Without tilt wheel
7805857	3-speed man.	Column	1967-68	Without tilt wheel
7805858	All	Floor	1967-68	Without tilt wheel
7804323	Three-speed auto	Column	1969	Without tilt wheel
7804324	3-speed man.	Column	1969	Without tilt wheel
7804325	All	Floor	1969	Without tilt wheel
7804365	Automatic	Column	1969	With tilt wheel
7804366	All	Floor	1969	With tilt wheel

Torque Specifications Rear Suspension

Upper Control Arm	Lower Control Arm	Shock Absorber	
Nut	Mounting Nut	Upper	Lower
80 lbs ft.	80 lbs ft.	60 lbs. ft.	20 -lbs ft.

Shock Absorber Bracket		Shock Absorber Nut
To Floor	To Mount	To Anchor Plate
8-lbs. ft.	8-lbs. ft.	40

Rear Spring		
Eye Bolt	Front Bracket	Rear Shackle
75-lbs ft.	25- lbs. ft.	50- lbs. ft.

Chapter 4

Wheels and Wheel Covers

Wheels

Several different wheels were used on the 1967-1969 Firebirds all of which are 14-inches in diameter. The standard width of the wheel was six-inches, it was listed as part number 9793125 all three years. The wheel was painted 60-degree gloss black (Dupont 88-L, Ditzler 9300, Rinshed-Mason P-403). This wheel was used with all models and options except with Rally wheels or disc brakes. A space saver spare, part number 9789327- 1967 and 9792737 1968-69, was optional, while the standard spare tire was a stamped steel unit with a conventional tire. Five lug nuts, part number 358501, these 7/16-in. nuts have a 20-thread count and should have dull silver appearance.

Some feel that mag styled wheels were required with the disc brake option, but that is not true. This stems from the fact that models with disc brakes required a special wheel, part number 9785955 ,with a larger off-set for the brakes to function properly. This wheel, or another approved wheel must be used with the disc brake option. The above wheel looks and is finished the same as the standard wheel, but it has a 1.00 inch off set. The reason for this mix up is that dealers misread the sales sheet and instead of thinking that styled wheels were available with the disc brake read them as that they were recommended.

Two different styles of styled wheels were available on the 1967 Firebird. First there is the limited produced Rally I wheel. There is confusion as to what this wheel looks like. Listed as part number 9781246 it looked much like a slotted Chevrolet Rallye wheel with a larger center (part number 9785501) cap and a bright (part number 9781480) trim ring. Each wheel was held on with five acorn shaped lug nuts listed as part number 9780732. The Rally I wheel should be painted argent silver. This wheel looked too Chevrolet, and did not find that many interested buyers.

Most buyers preferred the Rally II wheel to the Rally I wheel. The Rally II wheel came in two forms part number 9789329 without disc brakes and part number 9787279 with disc brakes. These five spoke wheels had strong features of a SS wheel, which was a very popular after market wheel back then. The inner spokes of this wheel were dark gray and the outer portions were argent silver. A PMD medallion part number 9787941 was placed in the center and the outer edge of the wheel was highlighted with a trim ring listed as part number 9781480. The top of the lug nuts part number 9789086 were filled in with red paint.

The Rally I wheel was discontinued at the end of the 1967 model year. The Rally II wheel continued unchanged into 1968.

In 1969 14x7 in wheels became available as the standard width on V-8 powered Firebirds, the 6-cylinder, except the 4-bbl version, still came standard with the 14x6 in. wheels, and used the same part numbers as it did in previous years. The Sprint six cylinders with a 4-bbl carburetor also used the wider 7 in. wheels, which were listed as part number 3966936. Likewise, with this increase in width came an increase to six inches for the space saver spare, listed as part number 9793592.

Rally II wheels were also increased in width, listed as part number 480723, they were available only on those models with a V-8 power plant, or on the Sprint optioned cars. On cars with tire chains or a 1-bbl six cylinder then the 14x6 inch Rally II wheels were used.

1967 Firebird Wheel Codes

Description	ID Code
Standard	HG
Disc Brakes Standard	HH
Rally I	KB
Rally II w/o Disc Brakes	JC
Rally II with Disc Brakes	JA
Space Saver	HE

1968 Firebird Wheel Codes

Description	ID Code
Standard	HG
Disc Brakes Standard	HK
Rally II w/o Disc Brakes	JC
Space Saver	HI

1969 Firebird Wheel Codes

Description	ID Code
Standard 6-cyld 1-bbl	HK
Standard V-8 or Sprint	IF
Rally II 6Cyld.	JC
Rally II V-8	JK
Space Saver 6Cyld.	HI
Space Saver V-8	KC

The optional space saver spare tire. Note the correct light gray color

The popular Rally II wheel. Shown is a 1967 model. Note the correct red line tires.

REFINISHING WHEELS

The standard stamped steel wheel, which was used with a full-wheel cover should be thoroughly cleaned and free of rust, then lightly sand the surface with 400-grit sandpaper. Then apply two thin coats of a durable enamel semi-gloss black paint. Do not use acrylic lacquer, as the surface will have to be buffed out, buffing out a wheel is hard to do with all the nooks and crannies that a wheel contains. Enamel provides a hard surface that wears well, plus does not require buffing to shine.

The standard 7/16-20 thread hex-head lug nuts part number 358501 should have a natural finish.

The Rally II wheels can be repainted this way (this information was taken from a 1975 dealer Installed Technical bulletin # 75-T-14 dated 10/27/75).

1. Clean the wheel thoroughly to remove all grease and dirt.
2. Thoroughly scuff sand total area to be refinished with # 400 grit sandpaper. If spokes do not need refinishing they can be masked at this time. If rust is present, sand rusted area to bare metal and clean the exposed metal with Prep-Sol or equivalent.
3. If paint is sanded through to bare metal spray area to be color coated with high quality dark gray surface-primer such as Ditzler DZL32 or equivalent. When dry, thoroughly sand the surface primer with 400 grit sandpaper to insure proper adhesion of color coat.
4. If bright argent spokes need to be repainted, spray with Ditzler DQ-8568. Allow to dry completely (author note: allow 12-24 hours).
5. Mask spokes as required. Scuff sand remainder of wheel to insure proper adhesion.
6. Spray charcoal gray areas (author note use Ditzler DDL32947)
7. Allow to dry (author note at least overnight) or bake at 180-degrees for 30-minutes.

Wheels Covers and Hubcaps

The standard cover on all models was a dog dish style hubcap listed as part number 9774609. This 10-1/2 in. diameter cap featured the words "Pontiac Motor Division" imprinted into the center and outlined with 42-ribs.

The optional wheel covers differed each year. However, there were basically three different options each year; the custom wheel cover, deluxe wheel cover, and the wire wheel covers.

For 1967 the custom wheel cover was listed as part number 9787697. This disk measured 15-13/32 in. diameter and featured 12 depressions, six of which were cut out with a tri-spinner with the PMD logo in the center. The Deluxe wheel cover was listed as part number 9714055; it measured 14-1/4 in. diameter and featured 26 holes and the PMD logo in the center cap.

The 1968 wheel covers were greatly changed, the custom cover was listed as part number 9791006. It featured eight round holes that gave it the illusion of a styled mag wheel. The deluxe cover was listed as part number 974122; it looked similar to the 1967 deluxe cover and has a similar red center with the black PMD letters.

For 1969 the custom cover was listed as part number 9795997. It was a simple looking cover that tried to mock the look of Rally II wheels, with five oval cutouts, flanked by five bright spokes, and a center dome. The deluxe cover, part number 9781045, had a similar design, but the spokes and cutouts were wider.

The third choice in wheel covers was the wire cover. Listed as part number 9781478 it was used all three years. This cover simulated the look of wire spoke wheels with the PMD logo in the center; this cover was available as a factory installed or dealer installed option.

Deluxe Wheel Covers 1967-1968 top 1969 bottom

1967 Firebird Tire Sizes

Model	Standard Tire	Optional
Except 400-ci	E70x14 BSW	E70x14 WSW E70x14 RSW 185Rx14 WSW
400-ci	E70x14 RSW	E70x14 WSW 185Rx14

BSW-Black side wall WSW- White Side wall
RSW- Redline tires R- Radial tire.

1968 Firebird Tire Sizes

Model	Standard Tire	Optional
6-cyld 1-bbl	E70x14 BSW	E70x14 WSW F70x14 BSW F70x14 WSW F70x14 RSW 195Rx14 WSW
6-Cyld 4-bbl	F70x14 BSW	F70x14 WSW F70x14 RSW 195Rx14 WSW
V-8 Except 400-ci	F70x14 BSW	F70x14 WSW F70x14 RSW 195Rx14 WSW
400-ci	F70x14 RSW	F70x14 WSW 195Rx14

BSW-Black side wall WSW- White Side wall
RSW- Redline tires R- Radial tire.

1969 Firebird Tire Sizes

Model	Standard Tire	Optional
6-cyld 1-bbl	E70x14 BSW#	E70x14 WSW F70x14 BSW F70x14 WSW F70x14 RSW 195Rx14 WSW
6-Cyld 4-bbl	F70x14 BSW	F70x14 WSW F70x14 RSW 195Rx14 WSW
V-8 Except 400-ci	F70x14 BSW	F70x14 WSW F70x14 RSW 195Rx14 WSW
400-ci	F70x14 RSW	F70x14 WSW 195Rx14

BSW-Black side wall WSW- White Side wall
RSW- Redline tires R- Radial tire.
#- Near end of model run thie F70x14 tire was replace with a F78x14 tire on this model.

Chapter 5

Drivetrain

Transmission, Three-Speed Manual

Standard equipment on all 1967-1969 Firebirds was a three-speed manual transmission, including the 1969 Trans Am, it is unclear that if any were built, if so, it would be only a handful. Shifter location depended on the model. For most the standard three-speed shifter was located on the column. The exceptions to this rule were those with the Sprint package and those with a 400-ci V-8, this includes the 1969 Trans Am, which used a floor shifted three-speed as standard affair.

1967

Most Firebirds that were equipped with a 3-speed manual transmission used the Saginaw built unit. Two different units were available. Part number 9789553 had a 2.85 first gear ratio and was used on all six-cylinders; this unit will have the code letters "FK painted in yellow on the right-hand side of the case. While those with all V-8's except the 400-ci used part number 3968036, it had a 2.54 first gear ratio and will have the code letters "RJ" on the right-hand side of the transmission case.

Column mounted shift lever part number 9790723 shifted both transmissions; this bright plated lever was topped off with a molded color knob that was keyed to the interior trim.

1967 Column Mounted Shift Lever Knobs

Color	Part Number	Interior Trim
Black	3899067	272
		269
		253
		259
Dark Blue	9777862	
Dark Turquoise	9777863	256
Med. Red	3899063	252
		258
Med Bright Blue	3899064	
Med. Gold	3899066	271
		267
		251
		257

A Warner heavy-duty three-speed manual transmission part number 3890533 was standard on models with a 400-ci V-8 in 1967. This unit incorporated helical gear throughout, that were specially designed to provide high torque capacity and gears that could operate at high speeds, without excessive heat build up or friction loss.

This transmission assembly used a floor mounted shift lever part number 3934284, with a black colored ball knob shifter part number 3921146. A floor shift lever was available as an option if the console was ordered. A dealer installed walnut knob part number 984989, was available for all floor shift transmissions in the Firebird.

1968-1969

Again all Firebirds except those with the 400-ci V-8 came standard with a Saginaw built unit. Again different units were available, according to engine family. Six cylinders still used a 2.85 first gear ratio while those with a V-8, except the 400-ci V-8 used a 2.54 first gear ratio. The type of shifter that was used also further divided transmissions; a different unit was used on one with a column shifterthan those with a floor shifter.

1968-1969
Saginaw Three-Speed Manual Transmissions

Engine Type	Shifter Location	Part Number
6-cyld.	Column	9793095
6-cyld.	Floor	9791989
V-8	Column	9793092
V-8	Floor	9791989

A bright plated column mounted shift lever was standard. The 1968 models used the same lever that was used in 1967; the 1969 models used part number 9796421, a black plastic knob part number 9792623 was used both years.

A floor mounted shift lever was available as an option RPO M09 or when the console was ordered. Listed as part number 9792435 this came with a black shift knob part number 9781840 was used both years. A unique option was the custom shift knob, when console was ordered then a wood grain knob part number 9794301 was used instead of the black knob.. This knob was also available as a separate option. A dealer installed option walnut knob part number 984849 was also available.

Models with a 400-ci V-8 came standard with a Dearborn built three-speed manual. Part number 9793425 was coded with the letters "DB" in yellow paint on the right-hand side of the transmission case. This transmission was standard with a floor mounted shift lever, which used the same lever and knobs that were used on the Saginaw unit, floor shift models.

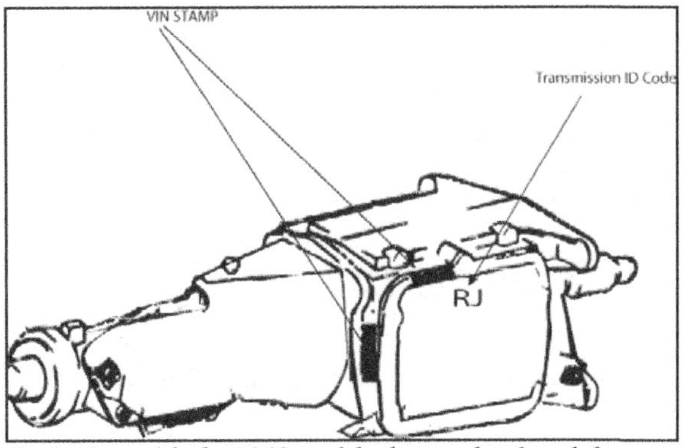

Beginning with the 1968 models the number 2 and the last eight digits of the cars VIN was stamped on the transmission case on Saginaw units it was located in either of these two locations. ID code was located on the R.H. side of the case.

Three-Speed Manual Transmission Identification Codes

Code	Shifter	Engine	Notes
		1967	
FK	Column	6-cyld.	
RJ	Column	V-8	Except 400-ci
ST	Floor	400-ci	H.D.
		1968	
FY	Column	6-cyld.	
FK	Floor	6-cyld.	Sprint
RJ	All	V-8	Except 400-ci
DB	Floor	400-ci	H.D.
		1969	
FK	Column	6-cyld.	
FY	Floor	6-cyld.	Sprint
RJ	All	V-8	Except 400-ci
DB	Floor	400-ci	H.D.

ll codes were on the right hand side of the case of all transmissions.

Beginning with the 1968 models a number that corresponded with the vehicles VIN was stamped into the case of all transmissions. The number consisted of the "2" for Pontiac followed by the last eight digits of that car's VIN. For your car to be considered a numbers matching car the number on the transmission must match those last eight digits in your car VIN that is located on the instrument panel.

REFINISHING TRANSMISSION

All three manufactuers used a cast iron main case and tail shaft. The transmissions were left unpainted when ne. To duplicate the fresh factory look of cast iron, after thoroughly cleaning the case and noting any paint marks (we will discuss this later on) the case assembly should be painted to cast iron gray.
Note that the Saginaw unit has a special finish, that is sort of a blue gray, which is the same color that was used on their steering gear boxes. The Ford built unit did not receive any finish and should be only painted cast iron gray to match the fresh cast appearance. The Warner transmission should be finished the same way as the Dearborn unit.

On Saginaw and the Dearborn transmissions code letters were used on the sides of the transmission for easy identification at the factory. These letters are about two-inches in height and appear on the right-hand side of the transmission case. The codes were stenciled on. You can find two-inch stencil lettering at an art supply or even a hardware store, the cardboard type stencils work excellently. And model spray paint is a close match to the color. On the left-hand side of the case you may find an inspector code. These are a two-digit code that are about 1-inch in height and appear in white near the rear edge of the main case. You may also find white marks that appear to be just a swab of paint on the units. These are not a mistake, they are line shift inspection marks indicating that the proper parts were put together. You will likely find one on the main case and one on the tail shaft. All shift rods should have a natural metal appearance.

Transmission, Four-Speed Manual

1967

Two different manufacturers offered four-speed manual transmissions for the Firebird line. The Muncie assembly plant built the majority of them. Available only on the V-8 powered Firebirds only one unit, part number 9789170, with a first gear ratio of 2.52 was available at first. This unit will have the code letters "FF" painted on top of the case. Later in the year a RPO M21 close ratio unit part 545769 with a 2.20 first gear ratio was offered, but only with 3.90 or 4.33 rear axle gears. This unit will have the code letters "FX" painted in yellow on the top of the case.

The Saginaw four-speed transmission assembly was also available, but only with the Sprint 6 model and by special request. Very few 4-speed Sprint models were built. No 4-speed 1-bbl cars were reportedly built. This unit, part number 9777732 has a 3.11 first gear ratio. The code "FH"(some have a "F7" code) can be found in yellow letters on the right-hand side of the transmission case.

Both the Muncie and the Saginaw unit were shifted manually with a floor mount lever listed as part number 9793456, with a black colored round shift knob listed as part number 3921147 that featured the shift pattern inlaid in white. Some used a white color knob.

1968

Both the Muncie and the Saginaw transmission return. The Muncie transmission was again available as a wide ratio part number 9792530 and the close-ratio unit part number 545769, yet this year did not have any axle restrictions.

The Saginaw transmission part number 9791992 was revamped and now sported 2.85 gears and was available on all 6-cylinders including the 1-bbl versions. Codes used on the transmissions were the same as those used in 1967.

Shift lever was slightly restyled and given a new part number of 9792437; the lever assembly was listed as part number 9792435 and was a Hurst built unit. It was used on both the Muncie and the Saginaw units. The knob was restyled and given part number 9794505.

Floor shifter without console, shown is the 1967 model.

Floor speed shifter with console, shown is a 1967 with a 400-ci powerplant

1969

The same Saginaw transmission that was used in 1968 was used again in 1969, and used the same code letters and part number. The Muncie transmissions were revamped and given new part numbers. The wide ratio was listed as part number 9798974 and the close ratio was listed as part number 9798975. Code letters were the same as they had been the previous two years.

Shift lever was again restyled and given a new part number of 9790770. The lever assembly however, used the same part number it did in 1968. Also using the same part number was the shift knob, a walnut knob with the shift pattern was also used. Listed as part number 9794300 it was used as a special ordered option.

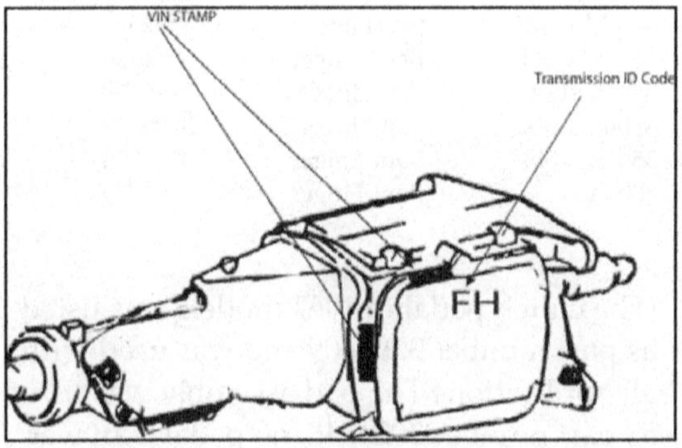

The Saginaw 4-speed has the ID code on the right-hand side of the case. The number 2 followed by the last eight digits of the car's VIN will appeared in either of these locations.

Saginaw Four-Speed Transmission Identification Codes

Code	Shifter	Engine	Notes
		1967	
FH	Floor	6-cyld	Special Order
		1968	
FH	Floor	6-cyld	Special Order
		1969	
FH	Floor	6-cyld	Special Order

ll codes were on the right hand side the casesmissions.

Muncie Four-Speed Transmission Identification Codes

Code	Shifter	Engine	Notes
		1967	
FF	Floor	V-8	Wide Ratio
FX	Floor	V-8	Close Ratio
		1968	
FF	Floor	V-8	Wide Ratio
FX	Floor	V-8	Close Ratio
		1969	
FF	Floor	V-8	Wide Ratio
FX	Floor	V-8	Close Ratio

ll codes were on the right hand side the call transmissions.

REFINISHING TRANSMISSION

The Saginaw transmission has the same finishing characteristics that were used on the three-speed manual transmission and should be followed. Except the height of the identification letters (FH), instead of being two inches in height as they are on 3-speeds the letters should be 1-inch in height. This is one of the most common mistakes that are being made in restoration of Firebirds. The change in height of the letters was done to quickly identify a four-speed from a three-speed at the factory.

The Muncie unit is different. The case was made of aluminum, and was left unfinished from the factory. To duplicate this finish and keep the unit looking new paint it with a dull aluminum silver finish such as Eastwoods Aluma Blast ® which very accurately duplicates the fresh look of smooth aluminum.

The code for the Muncie transmission is NOT located on the right-hand side of the case. Instead it was place on top of the case in 1-inch high letters. After applying the silver finish and the code letters it would be wise to cover it clear, to protect the finish and the codes.

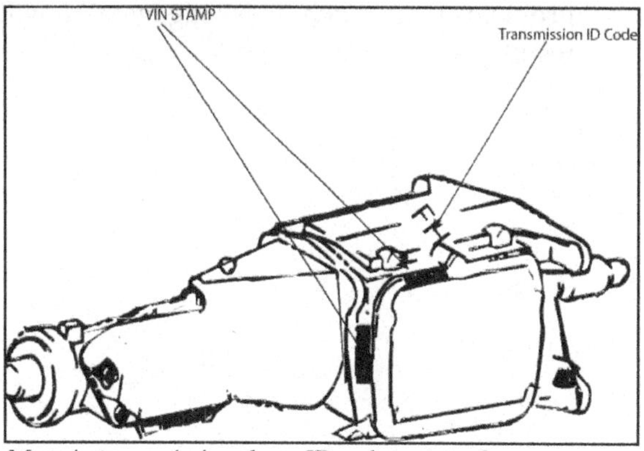

Muncie transmissions have ID code on top of case VIN stamp is in same location as other manual transmissions

Clutch

A diaphragm spring-type clutch is used on all Firebirds with a manual transmission. The clutch assembly consists of a drive plate, cover -pressure plate and release mechanism. Four different clutch driven plates were used, which depended on the engine size and output.

Clutch Driven Plates

Application	Diameter	Part Number
6-cyld 1-bbl	10.0 in.	9792928
6cyld 4-bbl	10.4 in.	9785970
326/350-ci	10.4 in	544185
400-ci	10.4 in.	9777438

The clutch cover and pressure plate of all models is the diaphragm spring type. Two types were used according to engine combinations. The flat finger diaphragm spring design was used on the 6-cyld 1-bbl and the 326-ci V-8s and the 1968-69 350-ci 2-bbl V-8s while the bent finger diaphragm spring design was used on 4-bbl powerplants, except the 1967 326-ci 4-bbl.

Clutch release bearing usage was determined on the clutch and pressure plate design. Those with the flat finger type will use a release bearing with an over all length of 1-7/8 in. and is stamped "CT-22" on the inner side of the bearing shoulder. While those using the bent finger type use a release bearing that is 1-1/4 in length and is stamped "CT-24K". Do not mix the two, as problems with clutch pedal engagement will results especially as the clutch wears.

Cover and Pressure Plate

Application	Type	Part Number
6-cyld 1-bbl	Flat Finger	3792398
6cyld. 4-bbl	Bent Finger	9781776
326-ci 2bbl	Flat Finger	3799139
350-ci 2-bbl	Flat Finger	3799139
350-ci 4-bbl	Bent Finger	9781776
400-ci	Bent Finger	9781776

The clutch pedal in 1967 models was listed as part number 3890609 and was used with all applications The pad assembly was listed as part number 3921648, no pedal trim was standard, but was available as part of a custom trim option. The trim plate was listed as part number 3923615.

For 1968 the pedal assembly was restyled and given part number 3965689. Although the pad assembly used the same part number as it did in 1967, the optional pedal trim was changed to part number 9777718.

Clutch pedal was again restyled in 1969 and will not fit earlier models. Listed as part number 9798209 the only thing it has in common with the 1968 version is the color. This pedal part number 9776819 and the pedal trim were available as part of an option package. It was listed as part number 9778267.

Originally the clutch pedals were dipped so the upper portion (that hidden by the instrument panel) was left unpainted. To be correct the upper portion of the pedal should have a natural appearance, which can be duplicated with paint, and the lower portion painted gloss black, the gloss depends on the model year. The 1967 models used a full 60-degrees gloss, while the 1968 and 1969 models used semi-gloss 30-degrees. A flattening agent that can be added to the paint can control the degree of gloss.

Transmission, Automatic

1967

There were two distinct types of automatic transmissions available for the 1967 Firebird line. The general type was a two-speed unit known as a Tempest automatic, so name because it was originally design for the mid-sized Pontiac line. The two-speed automatic transmission was available in all Firebirds except those with a 400-ci V-8.

A data plate that is located on the right-hand side of the transmission identifies the transmission. Transmission number, year and assembly date code will appear on the lower servo cover. The units used on six cylinders are air cooled while those used on V-8s are water-cooled. The Following transmissions and applications were used:

Two-Speed Automatic Transmission Codes

Application	ID Code	Part Number
6-cyld 1-bbl#	LF	1383471
6-cyld. ##	LH	1383748
6cyld 4-bbl#	LG	1383472
326 2-bbl	MA	1377113
326 HO 4-bbl	MB	1377114
326 ##	MC	1383315

#-w/o a/c ##with a/c

The 400-ci equipped birds came with a 3-speed Turbo-Hydra-Matic series 400 transmission when an automatic transmission was ordered. It was also optional as a heavy-duty unit on the 326-ci V-8 models.

The transmission is identified by a two-letter code that is stamped on the right-hand side of the case. Code letters and model year (proceeding a serial number) are stamped on the transmission nameplate, which is also located on the right-hand side of the case. The following transmissions and applications were available on the 1967 Firebird:

T.H. 400 Automatic Transmission Codes

Application	ID Code	Part Number
PV	326 2-bbl V-8	
PR	326-4-bbl V-8	
PX	400-ci#	9789929
PQ	400-ci Ram Air	9789928

except Ram Air

Standard shift lever for both types of transmissions was a column mounted brightly plated lever; part number 9790724 without tilt steering and part number 3895210 was used with tilt steering, with both type of transmissions. A plastic knob that was color-keyed to the interior trim was used on this shifter lever.

1967 Column Mounted Shift Lever Knobs

Color	Part Number	Interior Trim
Black	3899067	272
		269
		253
		259
Dark Blue	9777862	
Dark Turquoise	9777863	256
Med. Red	3899063	252
		258
Med Bright Blue	3899064	
Med. Gold	3899066	271
		267
		251
		257

When bucket seats and a console were both ordered then the shifter was moved to the floor, part number 9794098 was used in all applications. This shifter used a tee-shaped handle listed as part number 3895274 and a release button listed as part number 3914825.

1968

There were again two distinct types of automatic transmissions available for the 1968 Firebird line. The two-speed automatic was available on all powerplants except those with the 400-ci V-8. A data plate that is located on the right-hand side of the transmission identifies the transmission. Transmission number, year and assembly date code will appear on the lower servo cover. The following transmissions and applications were used:

Two-Speed Automatic Transmission Codes

Application	ID Code	Part Number
6-cyld 1-bbl#	LF	1383746
6-cyld. ##	LH	1383748
6cyld 4-bbl#	LG	1383747
350 2-bbl	MA	1383317
350 HO 4-bbl	MB	1383316
350 ##	MC	1383315

#-w/o a/c ##with a/c

The 400-ci equipped were the only Firebirds to come with the 400 Turbo-Hydra-Matic automatic transmissions. Very few 326-ci V-8s were ordered with the T.H. 400 transmission in 1967, and it was decided that it would not be offered for the 350-ci power plant in 1968.

Serial number of the transmission is located on the right-hand side of the case just forward of the governor. The serial number is preceded by the transmission identification code, which appears as two letters, this indicates what model and power plant the transmission was used with. The following codes were used in 1968 Firebirds:

T.H. 400 Automatic Transmission Codes

Application	ID Code	Part Number
PX	400-ci#	8626157
PQ	400-ci Ram Air	8626245
PY	400-ci with a/c	8626335

except Ram Air

Standard shift lever for both types of transmissions was a column mounted brightly plated lever; part number 9792626 without tilt steering and part number 9790725 was used with tilt steering with both type of transmissions. A black plastic knob part number 9792623 was used with all interior trim and transmission applications, except those with a console.

The floor shifter that was available when a console was ordered was completely restyled and will not fit earlier models. Listed as part number 9791314 it used a two part (part numbers 9791337-outer and 9791338-inner) tee-shaped handle.

1969

Saw the introduction of an all-new transmission. The Turbo-Hydra-Matic 350 was listed as option RPO M38. It was available on those with a six cylinder and the 350-ci 2-bbl V-8, except with air conditioning. This three-speed automatic was similar to the T.H. 400 but was more suitable to the lowered powered powerplants. In fact this transmission in later years would displace the two-speed automatic completely.

The production day and shift built number along with transmission identification code and model year are stamped on the 1-2 accumulator cover, which was located in the middle of the lower right-hand side of the transmission case. Don't confuse this with the VIN stamp which is located on the lower left-hand side of the case, next to the manual shift shaft. The following M38 transmissions were used on the Firebird.

M-38 Automatic Transmission Codes

Application	ID Code	Part Number
JB	6-cyld 1-bbl	6260053
JD	6-cyld- 4-bbl	6260055
JF	35-ci 2-bbl	6260003

The two-speed automatic was again available for select power plants on the 1969 Firebird. These included the 1-bbl 6-cylinders and the 350-ci 2-bbl, and with a 350-ci 2bbl and air conditioning. A data plate that is located on the right-hand side of the transmission identifies the transmission. Transmission number, year and assembly date code will appear on the lower servo cover. The Following transmissions and applications were used:

Two-Speed Automatic Transmission Codes

Application	ID Code	Part Number
6-cyld 1-bbl#	LF	1386505
6-cyld. ##	LH	1386507
350 2-bbl	MA	1386508
350 ##	MC	1386510

#-w/o a/c ##with a/c

Availability for the RPO M40 option Turbo--Hydra-Matic automatic transmission was revamped for 1969. It was again available, with a few select 350-ci models.

Serial number of the transmission is located on the right-hand side of the case just forward of the governor. The serial number is preceded by the transmission identification code, which appears as two letters, which indicates what model and power plant the transmission was used with. The following codes were used in 1968 Firebirds:

T.H. 400 Automatic Transmission Codes

Application	ID Code	Part Number
PX	400-ci#	8626517
PQ	400-ci Ram Air	8626516
PY	400-ci with a/c	
PV	350-ci a/c	8626520
PS	350-4-bbl	8626317

except Ram Air

Standard shift lever for both types of transmissions was a column mounted brightly plated lever; part number 9795118 without tilt steering and part number 9795890 was used with tilt steering with both type of transmissions. A black plastic knob part number 9792623 was used with all interior trim and transmission applications, except those with a console.

The floor shifter that was available when a console was ordered was listed as part number 9796736. A round knob part number 9793565 was used with all interior trims.

The Tempest automatic can be identified by a code located on the lower servo cover on the right-hand side of the transmission. Even though the MA code is correct for a 2-bbl V-8 Firebird the code "66" shows it is for a 1966 model and the code 026 means it was built on January 26 ,1966, making it for a Tempest model not a Firebird.

Automatic floor shifter, required the use of the optional console.

REFINISHING AUTOMATIC TRANSMISSIONS

The two-speed automatic case should have a natural look to it; painting it detail gray can duplicate this. While the Turbo-Hydra-Matic should have a brighter fresh aluminum appearance, painting the case aluminum silver can do this. White inspection marks were usually placed on the bell housing and the oil pan, these were just a dab of white paint and when reapplying do it in one quick motion. Also you may find chalk marks on the bell housing, that will be the code letters that identify the transmission.

Propeller Shaft

Usage was determined by transmission type. Overall length of each shaft and diameter remained the same throughout all three-years. However, there was a change in the design and the 1967 models with a manual transmission did use a special shaft. Propeller shafts should have a natural appearance or be painted flat black, they came both ways. They should also have a color code stripe on them. There was no set location for this code, most appeared in the middle of the shaft or near the tail-end of the shaft towards the rear axle. You may also find a dab of yellow paint on the shaft that will match up with a dab of paint on the axle pinion, this was done because the shaft is balanced and it is essential that the same location be used, if the shaft was ever removed.

Propeller Shaft Part Numbers and Color Codes

Transmission	Length	Diameter	Color Code	Part Number
3 and 4-spd Manual	49.96	2.75	2-Blue	7805686* 7805668+ 1968-69
2-speed Automatic	49.96	2.75	1-Pink	7801322- 1967 7805686- 1968-69
3-speed automatic	49.30	2.75	1-Red 1-Pink	7801899- 1967 7805687- 1968-69

*- Except Heavy Duty 3-speed +- Heavy Duty 3-speed

Rear Axle

A Salisbury type solid rear axle along with the leaf springs from a simple and efficient Hotchkiss drive system provided excellent ride and stable characteristics for a model at the time.

Axle housings differed each year, as did the axle gear ratios, the proper ratio can be determined by a code that is stamped on the left-hand axle tube of all axles. Note that gears were changed greatly in performance image cars like Firebirds, and the gears installed may not be the proper ones for your restoration. By counting the number of teeth on the ring and the pinion and dividing them, you can get the ratio of the gears. For example you have a ring with 39 teeth and a pinion with nine teeth you get a ratio of 4.33. Camaro axles are not interchangeable.

Rear Axle Part Numbers

Model Year	Part Number
1967	9785843
1968	9793293
1969	9797902
1969	546705*

*- Roller bearing

All Axle assemblies should be painted semi-gloss black for detail. A white sticker with red letter was applied to the left-hand brake drum with the appropriate rear axle code was also used. This sticker was quickly lost, and even sometimes removed in dealer prep of the car. The sole purpose of this sticker was for quick identification at the factory. The sticker was approximately 1-1/2 inches in height and width with red block letters that were approximately 1-1/4 inches in height. Other marks you may find on the axle will include white inspection marks especially over one of the rear cover bolts, which signified that the bolts had been torqued to proper specifications.

Several different differential cases were available each year. Part numbers depended on ratio, several different ratios could be used according to the fact that the same ring gear was used, but different pinion gears were used. The case was the same between standard and Safe-T-Track axles. A tag that is attached to the lower right hand section of the axle cover can identify safe-T-Track axles. There was a change in design in 1969 models early models will use 3/8x7/8- 24- right-hand thread ring gear bolts, while the later models will use 7/16x7/8 20- thread left-hand bolts, will usually be marked with the letter "L".

Differential Case with Gears Part Numbers

Part Number	Ratios
9784972*	3.08, 2.93, 3.23 Two-pinion case
9788828*	4.11, 3.36, 4.33, 3.90, 3.55 Four-pinion case
9784971*	2.79, 2.56 Two-Pinion case
479728#	3.36, 4.33, 3.90, 3.55 Four-pinion case
479726#	2.79, 2.56 Two-pinion case
479727#	3.08, 2.93, 3.23 Two-pinion case
546640*	3.90, 4.33 Four-pinion case

* 1967-early 1969 models #- Late 1969 models

1967 Firebird Axle Codes

Axle Code		Engine					Transmission				Ratio
Std.	Lock	6-cyl 1-bbl	6-yld 4-bbl	326-ci 2-bbl	326-ci 4-bbl	400-ci	3-spd Man.	4-spd Man.	AIR COOLED Auto	WATER COOLED Auto	
XB	UN	X		X					X	X	2.56
XC	UP		X						X		2.78
XD	UR	X		X						X	2.93
XE	US	X		X	X	X*	X	X		X	3.08
XF	UT	X	X	X	X	X*	X	X	X	X	3.23
XG	UV	X			x	X*	X	X			3.36
XH	UW		X			X*	X	X	X		3.55
	UX					X#	X	X			3.90

*-Except 400- Ram Air #- 400- Ram Air only

1968 Firebird Axle Codes

Axle Code		Engine					Transmission				Ratio
Std.	Lock	6-cyl 1-bbl	6-yld 4-bbl	350-ci 2-bbl	350-ci 4-bbl	400-ci	3-spd Man.	4-spd Man.	2-spd Auto	T.H. Auto	
XB	ZB	X		X					X		2.56
XC	ZC		X	X	X	X*			X	X	2.78
XD	ZD	X		X					X		2.93
XE	ZE	X		X		X*	X	X		X	3.08
XF	ZF	X	X	X	X	X$	X	X	X	X	3.23
XG	ZM	X		X	X	X*	X	X			3.36
XH	ZN		X		X	X*	X	X	X	X	3.55
	ZP				X	X	X	X		X	3.90
	ZR					X		X			4.33

*- Except Ram Air $- Except H.O. or Ram Air

1969 Firebird Axle Codes

Axle Code		Engine					Transmission				Ratio
Std.	Lock	6-cyl 1-bbl	6-yld 4-bbl	350-ci 2-bbl	350-ci 4-bbl	400-ci	3-spd Man.	4-spd Man.	2-spd Auto	T.H. Auto	
YB	ZB	X		X					X	X	2.56
YC	ZC		X	X		X*			X	X	2.78
YD	ZC			X					X	X	2.93
YE	ZE	X		X		X*	X	X		X	3.08
YF	ZF	X	X	X		X$	X	X	X	X	3.23
YG	ZG					X$	X	X			3.36
YH	ZH	X	X		X	X*	X	X	X	X	3.55
	ZK				X	X		X		X	3.90
	ZM					X		X			4.33

*- Except Ram Air $- Except H.O. or Ram Air

Chapter 6

Brakes

Master Cylinder

1967

Two different master cylinders were used in Firebirds. Usage depended on the brake type. All models with drum brakes, this include power and standard types, used part number 5458523 which used a 1.00-in. diameter piston and is stamped "BS" on the front end of the master cylinder. The master cylinder should be painted flat black, or have a natural unpainted appearance, which can be duplicated by painting it cast iron gray. The cover part number 5452320 should have a cadmium gold appearance and retainer part number 5468053 a natural look, painting with Testor's silver model paint can duplicate this. When disc brakes were ordered the master cylinder was listed as part number 5463492 which used a 1-1/8 in. diameter piston. Master cylinders used with disc brake applications were marked "WT" The natural appearing cover part number 5455517 was unique to the disc brake master cylinder only, and a cover from a master cylinder with drum brakes will not fit. Likewise the retainer part number 5455518 was also unique. Refinishing is the same as the standard brakes.

1968

Standard drum brakes again used a master cylinder with a 1.00-in. diameter piston. Now listed as part number 5458522, it was stamped "CT" and the cover and retainers are the same units that were used the year before. Those with disc brakes used part number 5459784 and was stamped "FR", cover and retainers were the same as those units used with disc brakes the year before. Refinishing is the same as the 1967 models, except more were left unpainted and have the cast iron gray appearance.

Brake master cylinders are identified by a code that is located on the end of the cylinder.

1969

Standard drum brakes used the same part number that was used in 1968. Those with disc brakes used part number 5469357 and was stamped "FR", other components were the same as the years before. Refinishing instruction are the same as previous years, see 1967 section for more details.

Brake lines should have a natural appearance; this is best achieved by paninting them detail gray in a semi-flat finish. The rear brake lines were routed along the left-hand frame rail supported by natural appearing clips and tap screws.

Booster Chamber

1967

Delco-Moraine built power booster chamber was used with power assisted drum brakes. It was listed as part number 5462049 and should be painted cadmium gold. The check valve was listed as part number 5462199. With disc brakes a Delco-Moraine unit was again used; it was listed as part number 5458828, it will not fit models with drum brakes. Finishing was the same as those with drum brakes.

1968

In an effort to save money on tooling cost the 1968 Firebird used the same booster chamber that was used on the Tempest line. Listed as part number 5463666 it ws used on cars with drum brakes. More and more Firebirds were being built with the optional disc brake set up which used part number 5468817; both of these units were built by the Delco-Moraine factory.

1969

Delco-Moraine built power booster chambers were used again but new part numbers appeared. The drum brake cars used part number 5469210, while the disc brake models used part number 5469345. These were unique to the Firebird line, and were not used on the Tempest models.

The mid—year Trans Am model required special heavy duty brakes. The booster chamber here is listed as part number 5469413, it was used only on the Trans Am with disc brakes. Finishing is the same as the 1967 models.

The booster assembly should be painted cadmium gold.

Brake Drums, Rotors and Calipers

1967

All Firebirds came standard with 9-1/2 in. diameter drum brakes at both the front and the rear. Steel drums with radial fins part number 9793677 at the front and 9788681 at the rear. The drums should be painted steel gray to duplicate the look of fresh forged steel.

Optional disc brakes displaced the front drums with 11 in diameter rotors and a hub assembly listed as part numbers 1382730, which would make this a universal General Motors parts. The disc was made of cast iron and should have a natural finish to it.

Be sure to mask this area off before attempting to refinish your rotors. The lug bolts should not have a cast iron finish, nor be painted, but be cleaned and give a light coat with a lubricant like WD40 to prevent rusting. For detail the center of the bolt can be painted, using a small brush, with detail gray to duplicate the look of fresh steel.

The calipers consisted of an outboard housing listed as part number 5456076, which fit either side, and an inboard side. The inboard housings were unique to each side. There were changes in the original pipe nuts and caliper inboard housings. The first type had 3/8-in diameter thread, while the second generation on the right-hand side had 7/16-in diameter threads and the driver's side used ½-in diameter threads. This change was done to avoid the possibility of installing the pipes on the wrong side or upside down. The first generation calipers were eliminated as replacement parts and the second type took their place. When replacing 1967 housing with later style the pipe (hose to caliper) must also be changed to the later style. If your Firebird still has the first type of housings then be sure that the contour of the pipe follows that of the caliper housing. The pipe should be installed with no bending; if any bending appears to be necessary then the pipe is either being installed upside down or on the wrong side of the car. (This information was taken from a dealer service bulletin number 67-1-61, there is no date of the change in the production on cars but this bulletin was dated April 21, 1967) .

Caliper Housing and Pipe Part Numbers

Part	Fist Type	Second Type
In board Caliper	5462042 LH 5462043-RH	5468363-LH 5468364-RH
Outboard Caliper	5456076	5456076
Brake Pipe	N/A	9789909-LH 9789908-RH

The caliper should have a natural appearance, which can be duplicated by painting it detail gray, be sure to protect the inner surfaces and rubber seals of the unit when painting. The backing plate was either natural, (duplicate by painting detail gray), or painted flat black, both styles were used.

1968

Very little was changed for 1968 , the rear brake drums used the same part number that was used in 1967, but the front drums were given new part numbers. Listed as part number 9790985 the rotor was still the two-piece design and used the same part numbers as it did in 1967, the calipers were the same units that were used on the later style 1967 models. All refinishing instructions are the same as they were in 1967.

There was a little mix up in the code error on the rear brake drums for some 1968 Firebirds. Certain early 1968 models using the Kelsey-Hayes built rear drums were marked with the identification code "F" instead of the correct "KH" code, this was due to switch over of "KH-F" as an optional stamping code. Note that some late built 1968 Firebird were reported to have used the 1969 front brake drums, which differed because of the fact that the drums could be service separately from the hub.

1969
Front drum brakes still measured 9-1/2 in. diameter and were given part number 9798690, these drums differed in the fact that they could be serviced separately from the hub. Rear drum brakes were the same part number as the previous two years. Big change was in the disc brakes; an all-new one-piece rotor listed as part number 405824 that still measured 11.00 in. diameter. Also restyled were the calipers which now used a single floating piston design. Listed as part numbers 5463806 left-hand and 5463807-right-hand, each side is unique, and will not swap sides. All refinishing was the same as in previous years.

Brake pad with manual transmission
Is the same cover that was used on the clutch pedal.

Brake Pedals

1967-1969
Brake pedal differed according to what type of transmission was used. Those with manual transmission used part number 3923541 for 1967 and 1968. The brake pad was listed as part number 3921648. No pedal trim was standard but was part of décor option group both years. The trim was different each year and listed as part number 3923615 in 1967 and 9777718 in 1968. For 1969 the pedal with manual transmission was listed as part number 9798202 and used pad part number 9776819. Again, pedal trim was part of an option package and used part number 9778267.

Those with automatic transmission used a brake pedal with a larger step pad. Two different pedal were used; 1967-68 models used part number 3923545 and the 1969 models used part number 9798200. Pedal pads followed the same format that was used on those with manual transmission; the pad was bigger to fit the larger pedal design. Part numbers were listed as: 3888682 for 1967-68 and 9774235 for 1969. Trim plates were not standard here either, but were available in the same packages that were used on those with manual transmissions. Trim plates were different each year, the following were used: 3888683-1967, 9777717- 1968, 9788316-1969.

Pedal finish depends on the model year. The 1967 pedals should be painted 60-degrees gloss black. In the 1968-69 models the pedal assemblies should be painted 30 degree-gloss black, note that some early 1968 models may still be painted 60-degees gloss black. Just like the clutch pedals the brake pedals were originally dipped in paint. This process left the top portion of the pedal unpainted. You can duplicate this look by painting the upper half to match fresh steel appearance.

Brake Pedal Paint Part Numbers

Model Year	Dupont	Ditzler	Rinshed-Mason
1967	88	9248	400-A0-946
1968-	99-L*	9266	107B40
1969	99-L*	9266	107B40

**-This number is for full gloss a flattening agent will have to be added to get the correct gloss .*
Different brands work differently so ask the sales clerk where you buy you paint for the proper mixing applications.

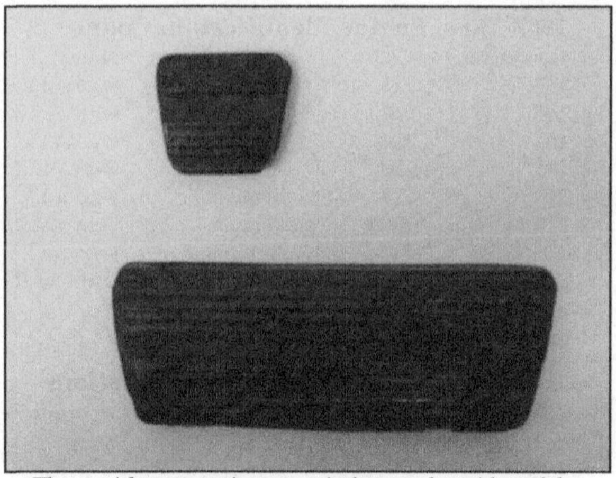

Those with automatic transmission used a wide pedal and pad, although the design is the same. The smaller pad above is for the parking brake.

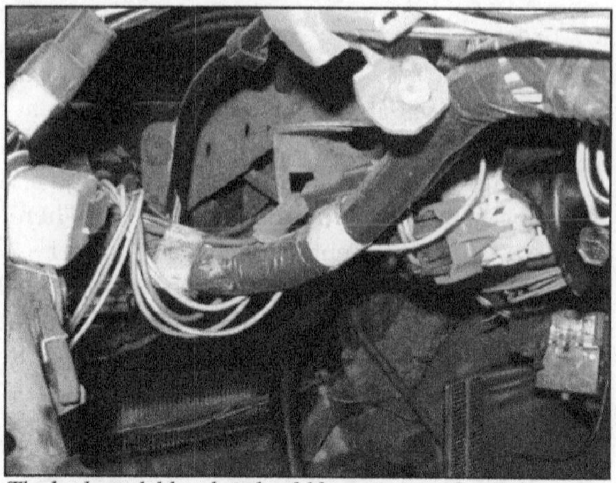

The brake pedal bracket should have a natural metal appearance.

Brake pedals were originally dipped in gloss black paint.

Parking Brake

1967-1969

Brake pedal differed according to what type of transmission was used. Those with manual transmission used part number 3923541 for 1967, and in 1968 the brake pad was listed as part number 3921648. No pedal trim was standard but was part of the décor option group both years. The trim was different each year and listed as part number 3923615 in 1967 and 9777718 in 1968. For 1969 the pedal with manual transmission was listed as part number 9798202 and used pad part number 9776819. Again pedal trim was part of an option package and used part number 9778267.

Chapter 7

Engines

Short Block, Six Cylinder

1967

Standard power plant for the base Firebird, but not available in any other model was a 230-ci overhead cam six-cylinder with a 9.0:1 compression ratio that breathed through a 1-bbl carburetor and pump out 165 hp. There were four different version of this power plant available. There was one with automatic and one with manual transmission without A.I.R. emissions, and one of each with A.I.R emissions.

A four-barrel version of this powerplant was available as standard equipment in the Firebird Sprint model; this powerplant added 10:5 compression, and a hotter cam along with the four-barrel carburetor that netted 215 horses. There were four versions of this power plant also. Same breakdown as above.

Both six cylinders used the same cast iron block. Individual engine identification is facilitated by a code that is stamped just below the engine production number. This code is stamped on the cylinder head contact surface of the block behind the oil filler pipe.

1967 230-ci Engine Identification Codes

Horsepower	Codes	Transmission	Notes
165	ZK	Manual	w/o A.I.R.
165	ZS	Manual	with A.I.R.
165	ZN	Automatic	w/o A.I.R.
165	ZM	Automatic	with A.I.R.
215	ZD	Manual	w/o A.I.R.
215	ZR	Manual	with A.I.R.
215	ZE	Automatic	w/o A.I.R.
215	ZL	Automatic	with A.I.R.

1967 230-ci Firebird Engine Specifications

Carburetor	Compression Ratio	Horsepower	Torque
1-bbl	9.0:1	165@ 4700	216@ 2600
4-bbl	10.0: 5	215@ 5200	240 @ 2800

1967 230-ci Short Block Part Numbers

Part	165-hp	215-hp
Cylinder Block	9788825	
Crank Shaft	9782279*	9782279*
Connecting Rod	3788435#	3788435 #
	9788924+	9788924+
Pistons	9777723	9777723
Piston Rings	9783357	9785695
Camshaft	9782012	9782218

*-Used bearing part # 3752487 with manual transmissions. #-Two reinforcements of caps +-one reinforcements on caps

1968

Standard six cylinder was bumped to 250-ci due to a longer engine stroke. This caused an increase of ten horsepower in the standard 1-bbl version, but oddly it did not effect the horsepower rating of the 4-bbl version. However, the torque rating was upped on both versions. Most likely the horsepower was raised on the 4-bbl six cylinder but due to marketing concerns the rating stayed the same. The actual rating was more like 230-hp. A total of four different six cylinder powerplants were available in 1968.

Both six cylinders used the same cast iron block. Individual engine identification is facilitated by a code that is stamped just below the engine production number. This code is stamped on the cylinder head contact surface of the block behind the oil filler pipe.

Starting this year the car's VIN was stamped on the engine block. The stamp will consist of the number "2" for Pontiac followed by the last eight digits of the VIN; a typical code would look like 28U120001, which would translate to the 20001st 1968 Firebird sold.

1968 250-ci Engine Identification Codes

Horsepower	Codes	Transmission	Notes
175	ZK	Manual	1-bbl
175	ZN	Automatic	1-bbl
215	ZD	Manual	4-bbl
215	ZE	Automatic	4-bbl

1968 250-ci Firebird Engine Specifications

Carburetor	Compression Ratio	Horsepower	Torque
1-bbl	9.0:1	175@ 4700	240@ 2600
4-bbl	10.0: 5	215@ 5200	255@ 3800

1968 250-ci Short Block Part Numbers

Part	165-hp	215-hp
Cylinder Block	9792278	
Crank Shaft	9790154*	9790154*
Connecting Rod	9793328	9793328
Pistons	9792414	9792414
Piston Rings	9792287	9785695
Camshaft	9790826	9792539

*-Used bearing part # 3752487 with manual transmissions

1969

This would be the last year for the overhead cam six cylinders, the 250-ci six cylinder would continue but in displacement only. Beginning in 1970 the design would be totally changed to a more common and more street manageable overhead valve set up. The little overhead cam six, was a strong performer, *Road And Track* magazine got a time of 17.2 seconds at 81 miles per hour in the quarter, and a 0-60 miles per hour time of 10.1 seconds. The test was done on a 1967 Firebird Sprint. However, the OHC style six cylinders suffered from problems such as the tendency to throw timing belts, and a inclination of the ignition jumping out of time, and most notably a problem with the throttle sticking wide open, because of insufficient throttle linkage clearance. Also due to their unconventional design, they were harder to work on and repairs were more costly. Nearly double over that of a conventional overhead valve type six cylinders.

Horsepower ratings were juggled strangely. The standard 1-bbl version was rated at 175 hp, in all forms. However, there were two different rating on the 4-bbl Sprint models. Those with a manual transmission were rated at 230-hp, while those with an automatic transmission kept the rating of 215 hp, which was very confusing to buyers.

A total of eight different versions were offered for this year. This was due in part to a change in design of the cylinder heads. Early style had 30-degree intake valve seat angles, while the later style used 45-degee intake valve seat angle. The short block did not change with this design, but different codes were used. The date of this change is not known, but for correctness later built Firebirds should use the later style codes. An early style block in late built Firebird would be considered incorrect.

Individual engine identification is facilitated by a code that is stamped just below the engine production number. This code is stamped on the cylinder head contact surface of the block behind the oil filler pipe.

As before in 1968 the block was also stamped with the VIN of the car. The stamp was located on the rear of the block on the right-hand side, behind the flywheel. The stamped consisted of the number "2" followed by the last eight digits of the car's VIN, but the entire VIN has appeared here before.

1969 250-ci Engine Identification Codes

Horsepower	Codes	Transmission	Notes
175	ZK	Manual	Early
175	ZC	Manual	Late
175	ZN	Automatic	Early
175	ZF	Automatic	Late
230	ZD	Manual	Early
230	ZH	Manual	Late
215	ZE	Automatic	Early
215	ZL	Automatic	Late

1968 250-ci Firebird Engine Specifications

Carburetor	Compression Ratio	Horsepower	Torque
1-bbl	9.0:1	175@ 4700	240@ 2600
4-bbl	10.0: 5	215@ 5200*	255@ 3800*
		230@ 5400**	260@ 3600**

*- Automatic Transmission **- Manual Transmission

1969 250-ci Short Block Part Numbers

Part	165-hp	215-hp
Cylinder Block	9798409	
Crank Shaft	9790154*	9790154*
Connecting Rod	9793328	9793328
Pistons	9796754	9796754
Piston Rings	9792287	9785695
Camshaft	9790826	9792539#

*-Used bearing part # 3752487 with manual transmissions
Stamped "E"

REFINISHING

All six cylinders blocks should be painted Pontiac Silver Blue (181-97219-DuPont, 13255, Ditzler, 266T21- Rinshed Mason) as should all other exterior engine components, this includes the oil pan. Two different oil pans were used: the 1967 models used part number 9782198, which is the same unit used on 1967 Tempests, while the 1968-69 models used part number 9791689, it too was used on 1968-69 Tempest models.

The timing chain cover was a two-piece plastic unit (upper- 9785713 and lower 978512), these units are molded in color. The crank case seal cover part number 9788802, should be painted to match the cylinder block. The timing belt part number 9782217 should have a natural black hard rubber appearance.

Short Block, V-8

1967

Standard V-8 the base Firebird, was 326-ci 2-bbl rated at 250-hp.. Next up and standard in the Firebird 326 H.O. series was a 326-ci 4-bbl rated at 285-hp. At the top end of the scale was the Firebird 400, which came in two styles the standard 400-ci 4-bbl, which rated at 325-hp, and a version with Ram Air that also used the same rating of 325-hp. Pontiac was a master of deception when it came to horsepower ratings. The rating for this power plant was taken at a lower rpm reading; the true full output of this power plant was more like 360 hp@ 5400 rpm. This under rating was done to not over shadow the GTO, and gives the Firebird an edge over the Camaro, when it was packing a 396-ci V-8. A Ram Air Firebird could out hustle a L78 396 Camaro even though the Camaro was advertised with more horsepower.

All Firebirds in 1967 with a V-8 came from the factory with a 326-ci or 400-ci power plant. Some dealers did install a 428-ci for the customers, but unless this can be documented in writing that it was installed before the customer took delivery of the car a 428-ci power plant in a 1967 Firebird would be considered incorrect.

Early models will have the casting number and date behind the distributor hole opening. On later models the casting number and date were moved to the rear of the block, just behind the number eight cylinder. The date consists of a letter code that indicates the month followed by the day and the year. Remember that the year may not always agree with the model year. Codes like D227 would be April 22, 1968 and before a 1967 model, while the code J257 would be September 25, 1968 and be for a 1968 model. Usually anything cast after July will be for the next year.

1967 V-8 Cylinder Block Casting Numbers

CID	Casting Number	Notes
326	9786339	2-bl/4bbl
400-ci	9786133	all

In a P &A Extra Bulletin Number 67-24 and dated October 3,1967 it reports that an all new block is available for Firebirds with 400-ci and Ram Air. This block, part number 9794075, was to be the replacement block for all cars with Ram Air. It is unique in the fact that it has 4-bolt mains. Is unclear if any cars from the factory were equipped from this block, as this bulletin was for the parts department. However, it is possible that this block was installed in cars before the owner tookpossession of the car, thus making a block with 4-bolt mains correct. Any car that contains this block, whether it was from the factory or dealer installed option the car will have to have to been built after October to be considered correct. It is believed that the block didn't actually appear in the part department till early November.

All V-8 blocks should be painted Pontiac Silver Blue 181-97219-DuPont, 13255, Ditzler, 266T21- Rinshed Mason. Oil pan 9778798 was used with all V-8 powerplants, it too should be painted to match the engine block. The dip stick was listed as part number 9777951 ,(unless air conditioning was ordered,) then part number 9789240 was used. The handle of the dipstick should be painted to match the cylinder block. The dipstick tube is also unique with air conditioning. Listed as part number 9789290, this tube has a curved shape, while the unit part number 523707 is straighter and used on cars without air conditioning. The tube should have a natural appearance, but a little engine color over spray is okay as the part was place on when the engine was painted.

1967 326-ci Engine Identification Codes

Horsepower	Codes	Transmission	Notes
250-hp	WC	Manual	w/o A.I.R.
250-hp	WH	Manual	with A.I.R.
250-hp	YJ	Automatic	w/o A.I.R.
250-hp	XL	Automatic	with A.I.R.
285-hp	WK	Manual	w/o A.I.R.
285-hp	WO	Manual	with A.I.R.
285-hp	YM	Automatic	w/o A.I.R.
285-hp	XO	Automatic	with A.I.R.

1967 400-ci Engine Identification Codes

Horsepower	Codes	Transmission	Notes
325-hp	WZ	Manual	w/o A.I.R.
325-hp	WU	Manual	with A.I.R.
325-hp	YT	Automatic	(1)
325-hp	WL	Automatic	Ram Air
325-hp	WQ	Manual	Ram Air w/A.I.R.
325-hp	XN	Manual	Ram Air (1)

(1)- Same code was used with or without A.I.R. applications

1967 326-ci Firebird Engine Specifications

Carburetor	Compression Ratio	Horsepower	Torque
2-bbl	9.2	250@ 4600	333 @ 2800
4-bbl	10.5	285@ 5000	359@ 3200

1967 400-ci Firebird Engine Specifications

Carburetor	Compression Ratio	Horsepower	Torque
Except Ram Air	10.75	325@ 4800	410@ 3400
Ram Air	10.75	325 @ 5200	410@ 3600

1967 326-ci Short Block Part Numbers

Part	250-hp	285-hp
Cylinder Block	9787915	9787915
Crank Shaft	9782770	9782770
Connecting Rod	541000	541000
Pistons	9783429	9783429
Piston Rings	9783358	9783358
Camshaft	9777254	9777254

1967 400-ci Short Block Part Numbers

Part	Except Ram Air	Ram Air
Cylinder Block	9787916	9794075*
Crank Shaft	9795480	9795480
Connecting Rod	541000	541000
Pistons	9788371	9788371
Piston Rings	9788376	9788376
Camshaft	9779067	9785744

1968

Standard V-8 the base Firebird, was now a 350-ci V-8, this is not a Chevrolet built powerplant it is completely Pontiac. Like the 326 the supplanted the 350-ci V-8 it came in two forms the standard version was a 2-bbl powerplant rated at 265-hp. Standard in the Firebird 350-HO was a four-barrel version of this powerplant that was rated at 320-hp. All in all the 350- 4bbl was a tried & true performer that offered all around good performance and good drive ability, with just a few modifications (free flowing air intake and exhaust system) it was easily capable of obtaining one horsepower per cubic inch

The 400-ci bruiser returned with a jump in horsepower but again they were under rated. The standard 400-ci 4-bbl was rated at 330-hp, the same exact power plant in a GTO was rated at 350-hp The reason for this underateing was the same as before corporate politics and the fact that they were trying to hide the power of this car from the insurance companies. Any car with the name Firebird was definitely on the hunting list for high insurance rates, and Firebird 400 packing 350 or over horsepower would be squarely in the cross hairs. This underatement also continued on to the two other 400-ci V-8s that were offered the H.O. rated at 335-hp and the Ram-Air version rated at the same rating as the GTO. Both of these powerplants were rated at 360-hp.

As the production year started only the Ram Air I option was available. On March 15, 1968 Ram Air II, as it is known today, was released for the Firebird 400 and GTO line. If your car was ordered before this date with Ram Air I, it was replaced with Ram Air II, as these were not separate options, but were simply listed as RPO L74 Ram Air. Any car built after March 15, 1968 would most likely have the Ram Air II package, but supplies of the Ram Air I had to be used up, so this may not be the exact date of the switch over.

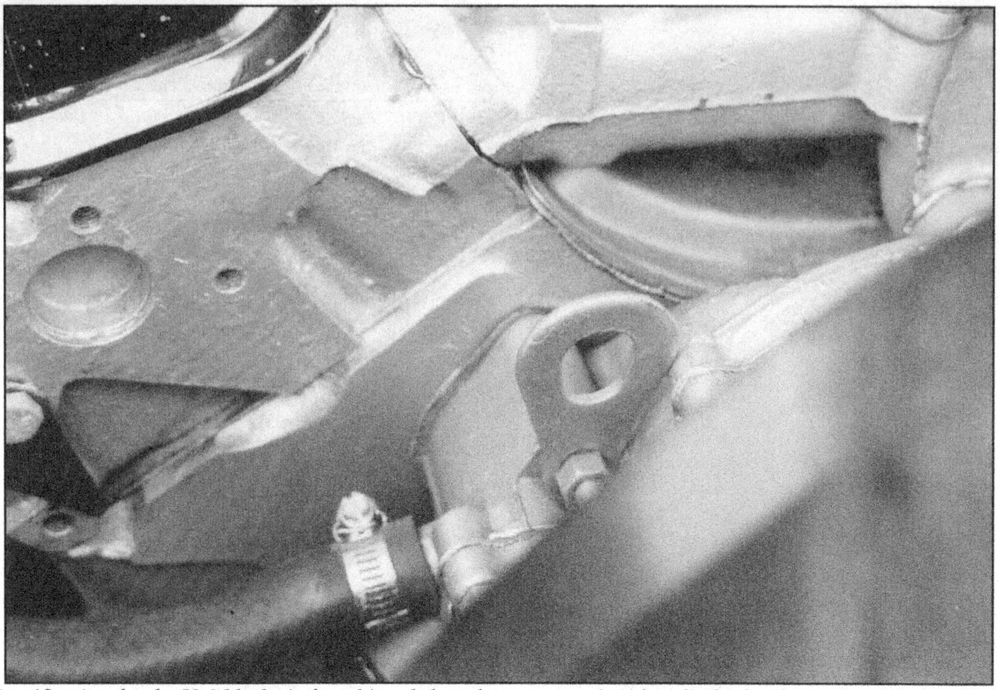

Identification for the V-8 blocks is found just below the passenger's side cylinder head. This code of YJ identifies it as a 1967 326-ci 2-bbl, not the 400-ci that the callouts on the hood said it was.

In the short block the Ram Air II package was a sign of the things to come, most of the parts were the same as those used on the 1969 Stage IV setup. This includes the four-bolt main bolt, camshaft forged pistons, and crankshaft. It also used special cylinder heads (see cylinder head section of this chapter), and special carburetors. The Ram Air II blocks used ID special code. It is reported that only 110 of these cars were built. Horsepower rating stayed the same, but it was more like rated at 370-hp, as this was what the Stage IV was rated at in 1969 in the GTO. Note: for your car to be correctly installed with the Ram Air II set up the casting date should be after the March 15, 1968 date. An early car with earlier casting date and or different ID codes would be considered incorrect.

The casting number and date are located at the rear of the block, just behind the number eight cylinder. The date used the same format that was used in 1967. Likewise the engine identification and production number are located in the same location as before, see the 1967 section for details.

1967 Casting Date Month Codes

Code	Month	Code	Month	Code	Month
A-	January	B	February	C	March
D	April	E	May	F	June
G	July	H	August	J	September
K	October	L	October	M	December

1968-1969 Casting Date Month Codes

Code	Month	Code	Month	Code	Month
A-	January	B	February	C	March
D	April	E	May	F	June
G	July	H	August	I	September
J	October	K	October	L	December

First generation Firebird engines could be decked out in chrome, liike this 326 H.O.

Beginning in 1968 all V-8 blocks were stamped with the number "2" indicating Pontiac and the last eight digits of the car's VIN, this is what is known as a matching numbers car. This stamping is located on a machined pad next to the timing chain cover. If the powerplant was changed during the warranty of the car, or by ordering a new engine from Pontiac then there will be no code on this pad.

All V-8 blocks should be painted Pontiac Silver Blue 181-97219-DuPont, 13255, Ditzler, 266T21- Rinshed Mason. Oil pan 9778798 was used with all V-8 powerplants, it too should be painted to match the engine block. The dip stick changed to part number 9793342 without air conditioning and part number 9793341 with air conditioning. When scourging for these parts look in the full-size 1968-1969 Pontiacs or the Tempest line, as they too used the same parts. These dipsticks were used in all V-8 models, regardless of the displacement. The handle should be painted to match the cylinder block. The oil dipstick tube was listed as part number 523707, which is the same unit that was used in 1967 on cars without air conditioning, and is used on most all other 1968 Pontiacs. The tube should have a natural appearance, but a little engine color over spray is as the part was placed on when the engine was painted.

1968 350-ci Engine Identification Codes

Horsepower	Codes	Transmission	Notes
265-hp	WC	Manual	2-bbl
265-hp	YJ	Automatic	2-bbl
320-hp	WK	Manual	4-bbl
320-hp	YM	Automatic	4-bbl

1968 350-ci Firebird Engine Specifications

Carburetor	Compression Ratio	Horsepower	Torque
2-bbl	9.2	265 @ 4600	355@ 2800
4-bbl	10.5	320@ 5100	380@ 3200

Dealer conversions to the massive 428-ci were not unheard of. However, for this to be considered correct paperwork must back up the claim and the conversion done by the original owner.

1968 400ci Engine Identification Codes

Horsepower	Codes	Transmission	Notes
330-hp	YT	Automatic	4-bbl
330-hp	YW	Automatic	4-bbl
335-hp	WQ	Manual	H.O.
335-hp	WZ	Manual	H.O.
335-hp	WI	Manual	Ram Air
335-hp	WZ	Automatic	Ram Air
335-hp	WU	Manual	Ram Air II
335-hp	WY	Automatic	Ram Air II

1968 400-ci Firebird Engine Specifications

Carburetor	Compression Ratio	Horsepower	Torque
Except H.O or Ram Air	10.75	330@ 4800	430@ 3400
H.O.	10.75	335@ 5000	430@ 3400
Ram Air	10.75	335@ 5300	430@ 3600

1968 350-ci Short Block Part Numbers

Part	265-hp	320-hp
Cylinder Block	9792279	9792279
Crank Shaft	9795479	9795479
Connecting Rod	541000	541000
Pistons	478809	478809
Piston Rings	9792421	9792422
Camshaft	9777254	9779066

1968 400-ci Short Block Part Numbers

Part	330-hp	H.O.	Ram Air
Cylinder Block	9792280	9794075	9794075* 9794192#
Crank Shaft	9795480	9795480	9795480* 9794054#
Connecting Rod	541000	541000	541000
Pistons	9788371	9788371	9788371 9794459
Piston Rings	9786464	9786464	9786464* 9788862#
Camshaft	9779067	9779068++ 9779067+	9785744++* 9779068+* 9794091#

*- Except Ram Air II #- Ram Air II
+-Automatic transmission ++_Manual transmission
Part # 9779068 marked "S"
Part # 9779067 marked "P"

Cylinder Block Casting Numbers			
Casting Number	**Displacement**	**Model Years**	**Notes**
9786339	326	1967	All types
9790079	350	1968-1969	All types
9786133	400	1967	2-bolt mains
9792506	400	1968	4-bolt Except Ram Air II
9790071	400	1969	2-bolt
9792506	400	1968-1969	1968 Ram Air II 1969 Ram Air IV

1969

Standard V-8 the base Firebird was again the 350-ci 2-bbl version and produced 265-hp. There are two different forms of this powerplant: the early models used the same format that was used in 1968, but the later style used different heads, and the results were given different identification codes. Your late built base Firebird V-8 should not have the early style ID codes, or vice versa. To do so would be considered incorrect.

The Firebird HO 350 was still standard with the 350-ci 4-bbl V-8, and was now producing 325-hp thanks in part to modifications to the short block that included a higher-lift, longer duration camshaft. Unlike it's 2-bbl sister this power plant was not changed during the year.

Firebird 400's came standard with a 400-ci 330-hp 4-bbl V-8. There were now two different Ram Air packages available. Order RPO L74 and you got the base Ram Air Package named Ram Air III, which was nothing like or as well equipped as last years Ram Air II. In fact, except for the air induction system and a few other parts this package was no different than the standard 400-ci although it was rated at 355-hp. Although the Ram Air II name did not continue most of the goodies did in the form of Ram Air IV. Listed as option code RPOL67, this package included the same special short block that was used on the Ram Air II models in 1968, but was rated at 345-hp.

The casting number and date, as well as the engine identification and production number are located in the same location as before, see the 1967 section for details. The VIN was again stamped on the block; see the 1968 section for more details on the location and format of this stamp.

All V-8 blocks should be painted Pontiac Silver Blue. Oil pan used the same part as it did in the previous years, as did the two dipsticks. Remember that there is a difference between those with and those without air conditioning, the one with air conditioning is longer. The oil dipstick tube was listed as part number 523707 for all powerplants except those with Ram Air IV which used part number 9795380, this tube was also used on 1968 models with Ram Air II. The tube should have a natural appearance, but a little engine color over spray is as the part was placed on when the engine was painted.

1969 350-ci Engine Identification Codes

Horsepower	Codes	Transmission	Notes
265-hp	WC	Manual	Early
265-hp	WM	Manual	Late
265-hp	XL	Automatic	Early
265-hp	XB	Automatic	Late
265-hp	YJ	Automatic*	Early
265-hp	YE	Automatic*	Late
325-hp	WN	Manual	4-bbl
325-hp	XC	Automatic	4-bbl

*-2-speed automatic with a/c 3-speed automatic T.H. 350 if without a/c

1969 400ci Engine Identification Codes

Horsepower	Codes	Transmission	Notes
330-hp	WZ	Manual	4-bbl
330-hp	YT	Automatic	4-bbl
335-hp	WQ	Manual	Ram Air III
335-hp	YW	Automatic	Ram Air III
345-hp	WH	Manual	Ram Air IV
345-hp	XN	Automatic	Ram Air IV

1969 350-ci Firebird Engine Specifications

Carburetor	Compression Ratio	Horsepower	Torque
2-bbl	9.2	265 @ 4600	355@ 2800
4-bbl	10.5	325@ 5100	380@ 3200

1969 400-ci Firebird Engine Specifications

Carburetor	Compression Ratio	Horsepower	Torque
4-bbl	10.75	330@ 4800	430@ 3300
Ram Air III	10.75	335@ 5000	430@ 3400
Ram Air IV	10.75	345@ 5400	430@ 3400

1969 350-ci Short Block Part Numbers

Part	265-hp	320-hp
Cylinder Block	9792279	9792279
Crank Shaft	9795479	9795479
Connecting Rod	541000	541000
Pistons	478809	478809
Piston Rings	9792421	9792422
Camshaft	9777254	9779066

1969 400-ci Short Block Part Numbers

Part	330-hp	Ram Air III	Ram Air IV
Cylinder Block	9792280	9792280	9794192
Crank Shaft	9795480	9795480	9794054
Connecting Rod	541000	541000	541000
Pistons	9788371	9788371	9794459
Piston Rings	9786464	9786464	9788862
Camshaft	9779067	9779068++ 9779067+	9794091#

+-Automatic transmission ++_Manual transmission
Part # 9779068 marked "S"
Part # 9779067 marked "P"

An example of a cylinder block casting number. The item at the left, is a casting clocking indicating it was cast at two- o'clock a shift indicator elsewhere will tell if it is a.m. or p.m.

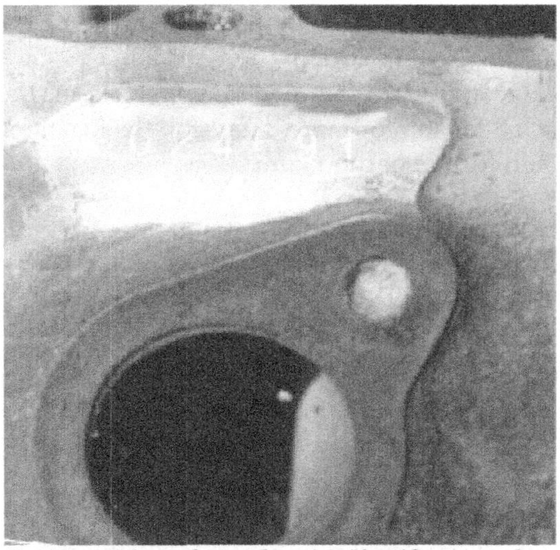

Engine serial number and engine identification code Are located at the front of the block, just below the right-hand cylinder head.

By the arrow code we can determine that this was cast by the day shift meaning that the clock in the photo above was 2 p.m. It was cast at GM-4 foundry.

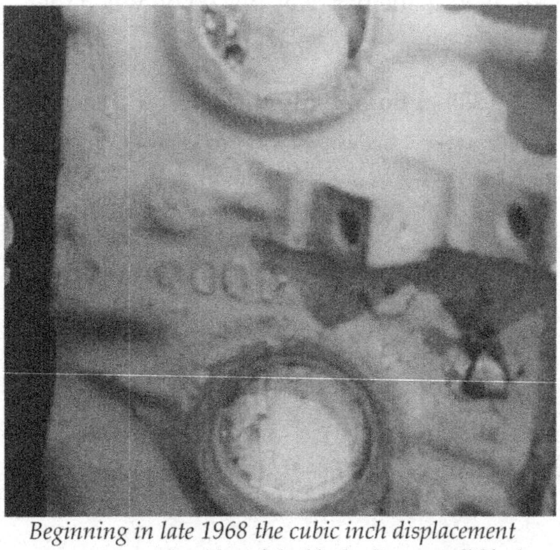

Beginning in late 1968 the cubic inch displacement was cast into the sides of the block. But not all blocks at all foundries were marked.

Cylinder Head, Six Cylinder

1967

The standard 1-bbl 230-ci six cylinder was a cast iron alloy unit listed as part number 9792641. This cylinder head used six (part number 9794218) steel intake valves. Six exhaust valves (part number 9791041). A ribbed valve cover listed as part number 9797513 was cast with the words "Overhead Cam".

The four-barrel version used a special cylinder head listed as part number 9792642. The head used intakes valves (part number 9782472) that were made of the same material and used the same head diameter as the 1-bbl version, but was longer. The same goes for the exhaust valves, which were listed as part number 9786805. No special valve cover was used with this power plant, as it used the same unit that was used on the 1-bbl version of the engine.

1967 230 Cylinder Head Casting Numbers

Type	Casting Number
1-bbl	9778692
4-bbl	9778693

1968

The standard 1-bbl 250-ci six cylinder again used a cast iron cylinder head. The cylinder head was restyled and given part number 484406. This cylinder head used six (part number 546376) 1.92-in diameter 1041 steel intake valves that were 4.81 in length, and six 1.60 in. head diameter 21-2N steel exhaust valves (part number 9791036) that were 4.80-in. length. A ribbed valve cover listed as part number 9797513 was cast with the words "Overhead Cam".

The four-barrel version used a special cylinder head listed as part number 9792262.. The head used intakes valves (part number 9794218) that were made of the same material and used the same head diameter as the 1-bbl version, but were longer at 4.90. The same goes for the exhaust valve, which was listed as part number 9791041, and were 4.89-in. length. No special valve cover was used with this power plant, as it used the same unit that was used on the 1-bbl version of the engine.

1968 250 Cylinder Head Casting Numbers

Type	Casting Number
1-bbl	9791194
4-bbl	9791195

1969

The standard 1-bbl 250-ci six cylinder began the year using the same exact cylinder heads that it used in 1968, with the 30-degree valve seats, all part numbers and casting numbers were the same as the 1968 models. Done as a rolling change, according to when supplies were used up, the cylinder heads were completely restyled. Part 9798597 used GM chrome intake valves listed as part number 9787330, although the dimension of the valve was the same this valve had a 45-degree angle seat The exhaust valve was not changed, and used the same part number it used in 1968. Also the same was the valve cover.

Early four-barrel 250-ci six-cylinders used the same heads that were used in 1968, including part numbers and the 1041 steel intake valves. During the model year these were changed to part number 9798598, which used GM 8840 steel intake valves listed as part number 9796651, like the 1-bbl version the exhaust valves did not change on this head.

1969 250 Cylinder Head Casting Numbers

Type	Casting Number
1-bbl	9791194- early/late
4-bbl	9791195- early N/A-late

Cylinder Head, V-8

1967

The standard 2-bbl 326-ci V-8 used a cast iron cylinder head listed as part number 9788062. It used eight steel intake valves part number 546408, and eight exhaust valves listed as part number 9787581. A pair of stamped steel valve covers part number 9786244- right hand and 9786245 left-hand. The covers were painted Pontiac Silver Blue to match the cylinder block in all models.

The 326-ci 4-bbl used a different set of heads listed as part number 9788063. These heads were also made of cast iron and each head used eight intake valves listed as part number 9778918, and eight exhaust valves listed as part number 9787578. The valve covers were the same exact items used on the 2-bbl set up.

Firebird 400 models used a cast iron intake listed as part number 9788067, which is the same head that was used on other Pontiac models with a 400-ci power plant. This cylinder head was used in all applications except those with ram air. Those with ram air used special heads that allowed the air to flow more freely; boosting horsepower, even thought it was not listed as doing so. Cylinder heads used on the Ram Air set up were listed as part number 9788067 for the first part of the year. These heads were replaced after engine serial number 646615. Where part number 9783657 took their place these heads feature push rod guides. Valve covers on the standard 400-ci engine were the same exact part that were used on the 326-ci V-8, while those used on the ram air packages used chrome plated valve covers listed as part numbers 9786315-right-hand and 9786316-left-hand.

1967 400-ci Valves and Valve Springs

Part	Intake	Exhaust	Notes
Except Ram Air			
Valves	9787587	9787584	
Springs	Inner	Outer	
	9779008	9779009	
Ram Air First Type			
Valves	9789725	9787584	
Springs	Inner	Outer	
	9779008	9779009	
Ram Air Second Type			
Valves	9792316	9792313	
Springs	Inner	Outer	
	9785999	9785998	

1967 326 Cylinder Head Casting Numbers

Type	Casting Number
2-bbl	140
4-bbl	141

1967 400 Cylinder Head Casting Numbers

Type	Casting Number
Except Ram Air	N?A
Ram Air Early	N/A
Ram Air Late	997

1968

With the introduction of the 350-ci powerplant new cylinder heads appeared. Those with the 2-bbl used part number 9792263. This cylinder head used SAE 1040 steel chrome intake valves that featured a 1.91 in. diameter head and were 5.02 in. length. The exhaust valves were 21-2N steel and used a 1.61 in. diameter head and were 5.01 in length. Valve cover was a painted stamped steel unit, and used the same part number as those used on the 1967 326-ci power plant.

Part number 9792264 was used on models with the 350-ci 4-bbl, also known to Pontiac as a H.O. unit. This head used GM 8440 steel intake valve, that were longer (5.12 in.) than the standard units. The exhaust valves were also longer (5.19 in.) but were made of the same material as the standard units.

Firebird 400 cylinder was again restyled and given part number 9790118. This cylinder head was the standard, non high performance unit and used the standard intake valves made of SAE 1040 steel chrome intake valves that featured a 1.91 in. diameter head and were 4.99 in. length. The exhaust valves used a 1.61 in. diameter head and were 4.98 in length. The ram air and H.O. units used a special cylinder head listed as part number 9792700, this head used the GM 8440 steel intake valves with a 2.06 in diameter head and were 5.08 in. length. The exhaust valves used a 1.72 in diameter head and were 5.07 in. length.

However, when Ram Air II replaced Ram Air I these cylinder heads were not used. Instead special high performance super duty heads part number 974040 were used. These heads featured round shaped ports instead of the D-port configuration with the siamese twin ports in the center, as in previous models. These heads were extremely free flowing and even though Pontiac didn't advertise it as such that was a huge booost in horsepower on these models, I would guess it at least 365-hp. Not only were the ports changed in this head, but also the push rods, pushrod guides and 1.50 in. rocker arms were used, and were serviced as a whole unit part number 9794325. This head also used special intake and exhaust valves, which are the same units that were used in the Ram Air IV package in 1969.

Valve covers were the same units that were used in 1967. The standard 400-ci 4-bbl used painted steel covers that were painted to Pontiac Silver Blue to match the engine block, while the Ram air models used chrome plated valve covers. See 1967 section for part numbers.

1968 350/400-ci Valves and Valve Springs

Part	Intake	Exhaust	Notes
350- ci			
Valves	9792326*	9792324*	*-2bbl
	9794311**	9793861**	**-4-bbl
Springs	Inner	Outer	All
	9779008	N/A	
400-ci Except Ram Air			
Valves	9792344	9792340	
Springs	Inner	Outer	
	9779008	9779009	
Ram Air I			
Valves	9792426	9792424	
Springs	Inner	Outer	
	9785999	9785998	
Ram Air II			
Valves	9794021	9794019	
Springs	Inner	Outer	
	9794044	9794045	

Standard style combustion chamber

High-performance large valve combustion chamber

1968 350 Cylinder Head Casting Numbers

Type	Casting Number
2-bbl	17
4-bbl	18

1968 400 Cylinder Head Casting Numbers

Type	Casting Number
Except Ram Air	62
Ram Air I	37
Ram Air II	96

1969

The 350-ci 2-bbl began the year using the same exact cylinder, including valves and springs that were used in 1968. Sometime during the year they were replaced with new head part number 9798599, this head was used only on engine codes WM, XB and YE. To use these heads on other engine codes would be considered incorrect. The intake valves were change in these heads from a 30-degree seat to a 45-degree seat. The valve covers were stamped steel units part numbers 9786244 right-hand and 9786245-left-hand were painted Pontiac Silver Blue and were used on both the early and later style heads.

The 350-ci 4-bbl used all new cylinder heads listed as part number 9795043, there was no mid year change for this head. This head used GM 8440 steel intake valves, which were longer (5.12 in.) than the standard units. The exhaust valves were also longer (5.19 in.) but were made of the same material as the standard units. There was a change in the valve springs. Early models used 9794934 inner and 9794939 for the outer, then after engine serial number 709185 part numbers 9781476-inner and 9779009-outer were used.

Standard Firebird 400 cylinder head was listed as part number 9790118 and this year all cylinder heads used the GM 8440 steel. Intake valves that featured a 1.91 in. diameter head were 4.99 in. length. The exhaust valves used a 1.61 in. diameter head and were 4.98 in length. This same head was also used on the Ram Air III package with automatic transmission. Those with manual transmission and Ram Air III used part number 9795043, which is the same exact head that was used on the 350-ci 4-bbl powerplant. Like the 350-ci 4-bbl valve springs were also changed on this model.

1969 400-ci Valve Springs

Type	Inner	Outer
Before Engine Serial Number 709186		
Ram Air III Manual Trans.	9794934	9794939
Ram Air III Automatic	9779008	9779009
Engine Serial Number 709186 and Up		
Ram Air III Manual Trans.	9781476	9779009
Ram Air III Automatic	9779008	9779009

The Ram Air III, also known as H.O., was a disappointing package when it came to brute horsepower, but was much more street savvy than other Ram air systems before it. For full out performance there was the 400 RAM Air IV, which used special heads listed as part number 979672. These heads were even freer flowing than the Ram Air II package, that it replaced as the hottest package available. This package offered the best of both worlds; it had the streetability of the Ram Air III package and the brut toughness of the Ram Air II package. This package used the same basic head design, with its round shaped ports, that were used on the Ram Air II, but employed 1.65 in. part number 5232675 with double locking nuts. A chrome plated valve cover was used on both ram air systems, however, it is common to find painted valve covers on those Ram Air III models. Stamped steel valve covers were used on all standard 400 models, the covers should be painted Pontiac Silver Blue to match the cylinder block.

1969 350/400-ci Valves and Valve Springs

Part	Intake	Exhaust	Notes
350- ci 2-bbl			
Valves	9792326* 9787325**	9792324	*-Early **-Late
Springs	Inner 9779008	Outer 9779009	All
350-ci 4-bbl/ 400-ci 4-bbl/ 400-ci Ram Air III auto.trans.			
Valves	9792344	9792340	
Springs	Inner 9779008	Outer 9779009	
Ram Air III Man. Trans.			
Valves	9792344	9792340	
Springs	Inner 9794934* 9779008**	Outer 9794939* 9779009**	* Early **-Late
Ram Air II			
Valves	9794021	9794019	
Springs	Inner 9796789	Outer 9796790	

1969 350 Cylinder Head Casting Numbers

Type	Casting Number
2-bbl	17- Early 47-Late
4-bbl	48

1969 400 Cylinder Head Casting Numbers

Type	Casting Number
Except Ram Air	62
Ram Air III	48- Manual 62- Automatic
Ram Air II	722

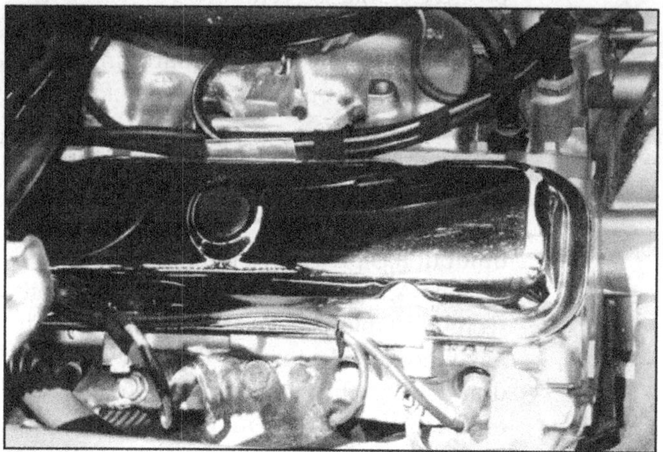

Chrome valve covers were used on cars with the H.O. packages, and Ram Air

Air Cleaner, Standard Six Cylinder

1967

Standard air cleaner assembly on the 230-ci 1-bbl six cylinders was single snorkel closed case design and was listed as part number 6484442. This air cleaner assembly should be painted 60-degrees gloss black. Those with A.I.R. emissions used a special air cleaner listed as part number 6483729, which featured provision for the A.I.R. emission connections The A.I.R emissions were only used in cars that were going to be registered in California. A heavy-duty version of both the air cleaners was available. Listed as part numbers 6483730 without air and without closed crankcase ventilation; 6483731 without A.I.R. with close crankcase ventilation and 6483732 with A.I.R., the heavy -duty air cleaner featured a replaceable oil wetted paper inner filter surrounded by a glycol wetted polyurethane foam outer filter.

The four-barrel version of the 230-ci six cylinder also used a closed single snorkel air cleaner. Standard in all states except California was part number 6483733, which was without A.I.R. emission or a closed crankcase ventilation system. A closed positive crankcase ventilation system was optional and this assembly part number 6483734 looked like the standard unit but had provisions on the under side of the base of the PCV hoses.

Standard in California was part number 6483735, which looked similar to the federal unit but has provisions for both the PCV and the A.I.R. emissions. Heavy-duty units were available as optional equipment, and have the same features as the heavy-duty unit used on the 1-bbl version above. The following heavy-duty air cleaners were available on the 230-ci 4-bbl:

1967 230-ci 4-bbl Heavy Duty Air Cleaners

W/O A.I.R. or PCV	6483136
W/O A.I.R. with PCV	6483737
With A.I.R.	6483738

1968

Air cleaner usage was much more simple for the 250-ci six cylinder in 1968. The 1-bbl version used only one air cleaner assembly. Part number 6483956 was used in all states including California. Painting instructions are the same as in 1967. The 4-bbl version used part number 6483958. The heavy duty air cleaner option was deleted.

1969

The air cleaner assembly was changed to part number 6485473 for the 1-bbl version, the 4-bbl version remained unchanged and used the same part number as in 1968.

Air Cleaner, Standard V-8

1967

A host of air cleaners were available on the 326-ci 2-bbl V-8, this was due to the fact that A.I.R. and PCV emissions effecting the usage, transmission type also effect the air cleaner assembly. A total of 11 different air cleaners were available. All air cleaners should be painted 60-degree gloss black. The following air cleaners were used on the 1967 326-ci 2-bbl Firebird:

1967 326-ci 2-bbl Air Cleaners-Federal

Application	Transmission	Part Number
	STANDARD	
W/O P.C.V	Manual	6424882
W/O/ PCV	Automatic	6424793
With PCV	Manual	6424883
With PCV	Automatic	6424794
	HEAVY-DUTY	
W/O P.C.V	Manual	6424885
W/O/ PCV	Automatic	6424796
With PCV	Manual	6424886
With PCV	Automatic	6424797

1967 326-ci 2-bbl Air Cleaners-California

Application	Transmission	Part Number
	STANDARD	
With A.I.R	Manual	6424884
With A.I.R.	Automatic	6424795
	HEAVY-DUTY	
With A.I.R.	Manual	6424887
With A.I.R.	Automatic	6424798

Four-Barrel version or the 326 HO powerplant also used a host of air cleaners but only six were available as transmission type did not effect the usage of the air cleaners on this Firebird. Air cleaners should be painted 60-degree gloss black. The following air cleaners were available:

1967 326-ci 4-bbl Standard

W/O A.I.R. or PCV	6424789
W/O A.I.R. with PCV	6424790
With A.I.R.	6484472

1967 326-ci 4-bbl Heavy Duty

W/O A.I.R. or PCV	6424791
W/O A.I.R. with PCV	6424792
With A.I.R.	6424505

The standard set up in the 1967 Firebird 400 was part number 6484473 this unit has a chrome cover and is used with a PCV set up. An air cleaner without the PCV package was optional and was listed as part number 6482590, it too used a chrome cover. Those Firebirds being shipped to California used a special air cleaner assembly listed as part number 6423211. Heavy-duty air cleaners were optional at extra cost. Part number 6424575 was used without A.I.R. emission and with PCV, and part number 6423209 was used with the A.I. R. emissions. Ram Air is a complicated option and is discussed in a separate section.

1968-1969

Choices for the 350-ci 2-bbl were narrowed to just two different air cleaner assemblies for the 1968 and 1969 model year. Part number 6424820 was used on cars without A.I.R emissions while part number 6424821 was used with A.I.R. emissions. The 4-bbl version of the 350-ci Firebird used only one air cleaner listed as part number 6485501 both years.

The standard air cleaner used in the 1968 and 1969 Firebird 400 was listed as part number 6424827, which is the same air cleaner assembly that was used on the full-size Pontiac with a 428-ci and the 1968 GTO. The 1969 GTO used part number 6484994, there is evidence that this air cleaner assembly was used on the 1969 Firebird. The air cleaner design was changed with a snorkel design; the base was painted 60-degrees gloss black, the lid was chrome.

A preheater assembly was used both years. A semi-flat black heat shroud 9791566-1968 and 9798815 1969 was used over the left-hand exhaust manifold and a tube or duct (part number 9793111- 1968 and 9791541-1969) was routed from here to the snorkel on the air cleaner assembly. These parts can be very hard to find. The 1968 parts are for Firebird models only, while the 1969 parts can also be found on the full-size Pontiacs with a 4-bbl V-8. All above part numbers are for models without Ram Air.

Chrome plate air cleaner used on 1967 H.O. models.

Carburetor, Standard 1-bbl

1967

A Rochester BV series single barrel downdraft carburetor was standard on the 230-ci six cylinder. Four different carburetors were used. Transmission and A.I.R. emission affect their usage. Identification number appears on small blue color tag that is attached to the top of the air horn. The carburetor has a natural finish. The following carburetors were used :

1967 230-ci 1-bbl Carburetor Part Numbers

Application	Transmission	Part Number
Without A.I.R.	Manual	7027167
Without A.I.R.	Automatic	7027168
With A.I.R.	Manual	7037167
With A.I.R.	Autoamtic	7037168

Throttle cable was listed as part number 9789016, and was used with all 1-bbl carburetors. Throttle return spring part number 9787064 should have a flat black appearance, and was color coded on the end with a dab of white paint. Throttle cable bracket was listed as part number 9786185.

1968

A Rochester BV series single barrel downdraft carburetor was used on the 250-ci six cylinder. Because A.I.R. emission package was deleted in 1968, only two carburetors were used. Part number 7028075 with manual transmissions and 7028168 with automatic transmissions. Identification number appears on small blue color tag that is attached to the top of the air horn. The carburetor has a natural finish.

Throttle cable was listed as part number 9792222, and was used with all 1-bbl carburetors. Throttle return spring part number 9792018 ws used with manual transmissions and part number 9787064 was used with automatic transmissions. The springs should have a flat black appearance, and was color-coded on the end with a dab of green paint for the automatic and white for the manual. Throttle cable bracket was listed as part number 9792202.

1969

The single barrel carburetor was still made by Rochester but was now a MV series. A total of four different carburetors were used. Those models with air conditioning used a special carburetor

1969 350-ci 1-bbl Carburetor Part Numbers

Application	Transmission	Part Number
Without A/C	Manual	7029165
Without A/C	Automatic	7029166
With A/C	Manual	7029167
With A/C	Autoamtic	7029168

Throttle cable, bracket and return spring remained the same and used the same parts that were used in 1968.

Typical Firebird air cleaner for non-H.O. powerplants, this is the heavy-duty unit for a 1968 350-ci V-8.

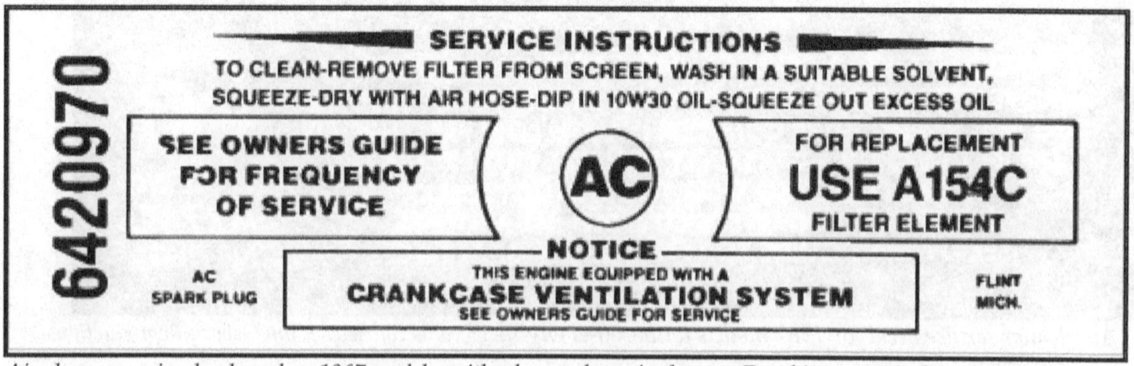

Air cleaner service decal used on 1967 models, with a heavy-duty air cleaner. Decal is courtesy of Jim Osborn Reproductions.

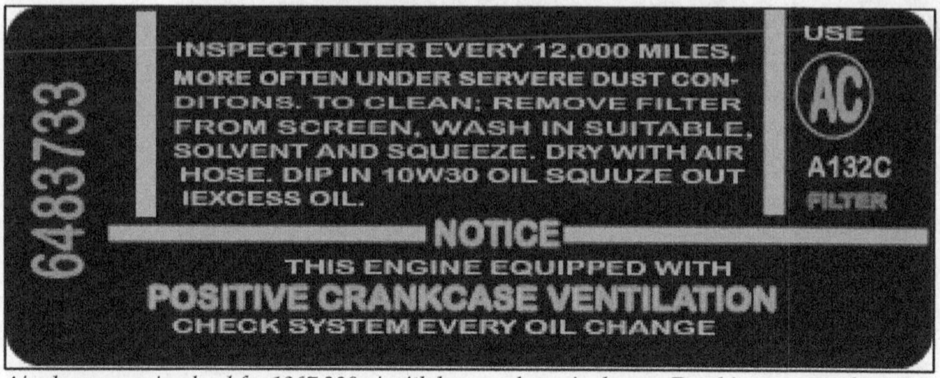

Air cleaner service decal for 1967 230-ci with heavy –duty air cleaner. Decal is courtesy of Jim Osborn Reproductions.

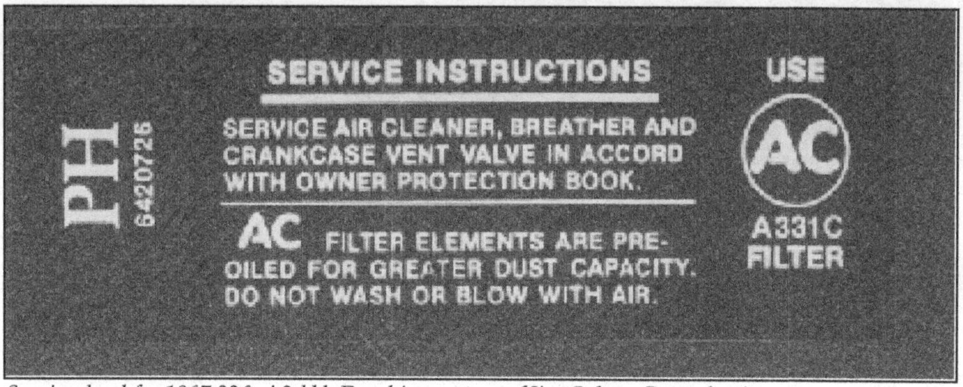

Service decal for 1967 326-ci 2-bbl. Decal is courtesy of Jim Osborn Reproductions.

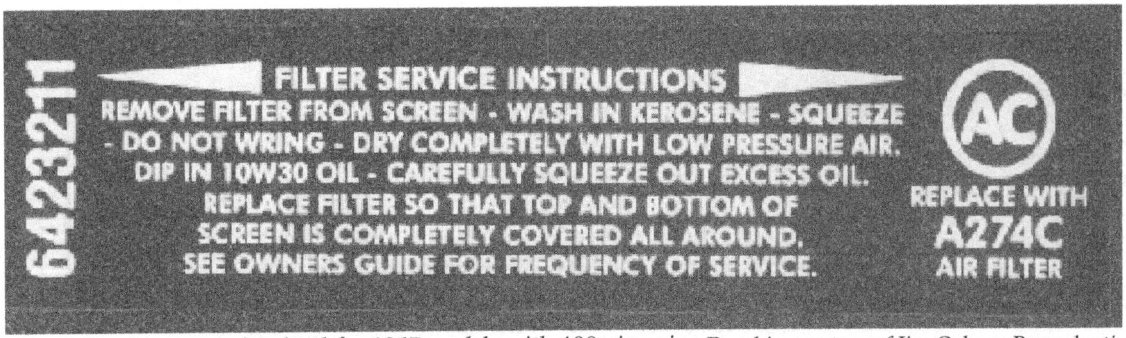

Air cleaner service decal for 1967 models with 400-ci engine Decal is courtesy of Jim Osborn Reproductions.

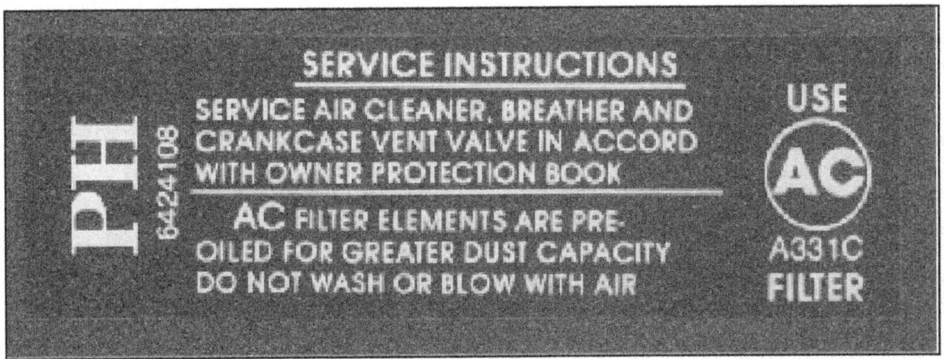

Service decal for 1968 350-ci 2-bbl V-8. Decal is courtesy of Jim Osborn Reproductions.

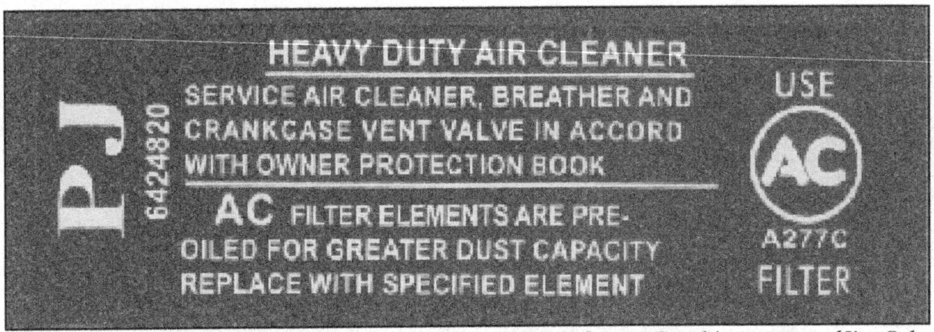

Air cleaner service decal for 1968 350-ci with heavy duty air cleaner. Decal is courtesy of Jim Osborn Reproductions.

Ram Air 400 required a special air cleaner set up that allowed cold outside air to flow down through the hood scoops through a special trap door on the air cleaner and then into the carburetor, thus increasing horsepower.

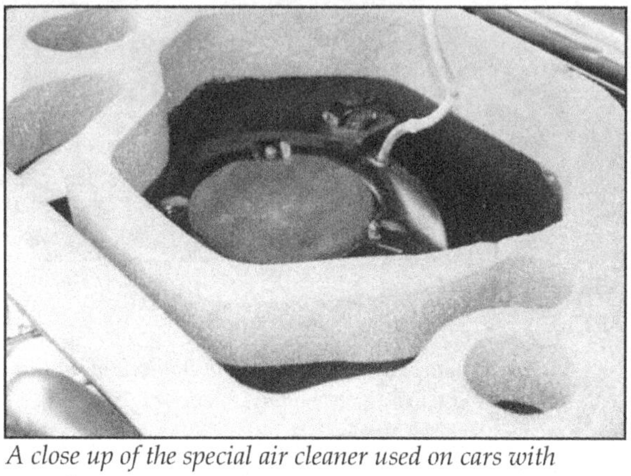

A close up of the special air cleaner used on cars with ram air.

Service decal used on 1968-1969 air cleaners.
Decal courtesy of Jim Osborn Reproductions.

Carburetor, Standard 2-bbl

1967

A Rochester 2GC series 2-bbl carburetor was used on the standard Firebird V-8. A total of four different carburetors were used. Those in all states except California used part numbers 7027071 with manual transmissions and 7027062 with automatic transmissions. Cars being shipped to California required special carburetors that has provisions for the A.I.R. emissions. These carburetors were listed as part numbers 7037071 with manual and 7037062 with automatic transmissions.

Throttle cable was listed as part number 9783230 it was 11-1/2 in length and coded with red tape. The throttle return spring part number 9792018 was used with manual transmissions and part number 9787064 was used with automatic transmissions. Throttle cable bracket was listed as part number 9787161.

1968

A Rochester 2GV series 2-bbl carburetor was used on the standard Firebird V-8. Due to the deletion of the A.I.R. emission package only two carburetors were used. Part number 7028071 was used on cars with a manual transmission and part number 7028062 with an automatic. The carburetor part number will appear on a small tag that is attached to the top of the air horn. Also the part number was cast into the body of some carburetors.

Throttle cable was listed as part number 9792222, which is the same cable that was used on the 1-bbl six cylinder. The throttle return spring was listed as part number 9789460, it should have a flat black finish and was color coded at the end with yellow paint. Throttle cable bracket was listed as part number 9797241.

1968

A Rochester 2GV series 2-bbl carburetor was again used on the 350-ci standard V-8, in the base Firebird only. Part number 7029071 was used on cars with a manual transmission and part number 7029062 with an automatic. The carburetor part number will appear on a small tag that is attached to the top of the air horn. Also the part number was cast into the body of some carburetors.

Throttle cable was listed as part number 9792222, which is the same cable that was used on the 1-bbl six cylinder. The throttle return spring was listed as part number 9789460, it should have a flat black finish and was color coded at the end with yellow paint. Throttle cable bracket was listed as part number 9797241.

Another view of a 1968 model with 350-ci v-8, note the painted valve covers, which is correct for this model.

Carburetor, 4-bbl-H.O.

1967

Three different engines were available with a four-barrel carburetor. The 230-ci six-cylinder that was standard in the Firebird Sprint model used a Rochester model 4MV. Two different carburetors were used, according to transmission type. Part number 7027261 with manual transmission and part number 7027260 was used in all states except California. Those Firebird Sprints being shipped for registration in California used different carburetors that were made for use with the A.I.R. emissions package. Those with manual transmission used part number 7037261 and those with automatic used part number 7037260.

The identification and change code is stamped on the throttle lever side of the bowl housing.

Throttle control cable was listed as part number 9783228 is 14-1/4 in. length and is color-coded with blue tape. The throttle return spring was listed as part number 9783282 it should have a flat black appearance and color-code on the end with flat pink paint. The throttle control bracket part number 9786682 should be painted to match the engine block.

The Firebird 326 HO used a Carter AFB 4-bbl carburetor. The body of this unit was made from aluminum and it should have natural aluminum appearance. The identification tag was located on the rear of the air horn's right-center screw. Four different carburetors were used according to the transmission type and emissions. Firebird 326 HOs registered as a new car in California used special carburetors that are different than the other 49 states.

1967 326 HO Carburetors

Without A.I.R.	Manual	AFB4243
Without A.I.R	Automatic	AFB4246
With A.I.R.	Manual	AFB4245
With A.I.R.	Automatic	AFB4248

The Firebird 400, except those with Ram Air, used a Rochester model MV four-barrel carburetor. Like other units transmission type and emission packages affected carburetor usage, the difference here was no special carburetor was used in California with the manual transmission, It is important to understand that we are discussing those models without Ram Air, as these models used a different seat of carburetor and will be covered in a separation section entitled Ram Air. Identification is the same as those on the six-cylinder models.

1967 400 HO Carburetors Without Ram Air

Without A.I.R.	Manual	7027273
Without A.I.R	Automatic	7027272
With A.I.R.	Automatic	7027272

Throttle cable on the 326 HO was the same unit that was used on the 2-bbl version. The cable on the Firebird 400 was listed as part number 9788674. Throttle return spring was the same unit that was used on the 2-bbl V-8 models, and was used for both the 326 and 400-ci models .Throttle control bracket was listed as part number 9787162 for 326 HO Firebirds and 9792069 for those with the 400-ci.

1968

All H.O powerplants this year used a Rochester model 4MV carburetor. With the cylinder heads redesigned to provide better emission standards, no A.I.R. package was required in California cars. However, transmission type did affect the usage of the carburetor.

The identification and change code is stamped on the throttle lever side of the bowl housing.

1968 Firebird HO Carburetors

250-ci	Manual	7028261
250-ci	Automatic	7028260
350-ci	Manual	7028269
350-ci	Automatic	7028266
400-ci	Manual	7028265
400-ci	Automatic	7028264

Throttle cable on the 250-ci 4-bbl was listed as part number 979223, while the 350 HO and 400-ci, without Ram Air, used part number 979224. The throttle return spring for the six-cylinder was listed as part number 9783282, while both the V-8 4-bbl used part number 9789460. Throttle cable mounting bracket was listed as part number 9792233 for the six-cylinder and part number 9792242 was used on the V-8s. Both should be painted the same color as the engine block.

1969

Again all four barrels were Rochester 4MV Quardajets. The identification and change code is stamped on the throttle lever side of the bowl housing. Transmission type again affected the usage of the carburetor.

1969 Firebird HO Carburetors

250-ci	Manual	7029261
250-ci	Automatic	7029260
350/400-ci	Manual	7029263
350/400-ci	Automatic	7029268

Throttle cable on the 250-ci 4-bbl was listed as part number 979223, while the 350 HO and 400-ci, without Ram Air, used part number 979224. The throttle return spring for the six-cylinder was listed as part number 9783282, while both the V-8 4-bbl used part number 9789460. Throttle cable mounting bracket was listed as part number 9796618 for the six-cylinder and part number 9797415 was used on the V-8s. Both should be painted the same color as the engine block.

Intake

1967

All intakes were painted Pontiac Silver-Blue to match the engine block.The casting date on the intake manifold is one of the most important features on the unit. Care must be used to ensure that the proper casting date is used. The casting date usually appears in the same general location that the casting number was, (on V-8's) it was located between the temperature gauge sending unit and the waterneck on the left-hand runner. The casting date breaks down the same as the engine block, and uses the same codes. To be proper the casting date must be BEFORE the assembly date for your Firebird. Typically the date should be any where from two to six weeks before the assembly date.

1967 Firebird Intake Manifold Part Number

Carburetor	Part Number	Notes
	230-ci Six Cylinder	
1-bbl	9794231	
4-bbl	9787541	
	326-ci V-8	
2-bbl	9782893	
4-bbl	9782895	
	400-ci V-8	
4-bbl	9786285	

1968

All Firebird intakes were painted Pontiac Silver-Blue to match the engine block the casting date for V-8 models is usually located on the front water passage just behind the waterneck. The water neck for all V-8s was listed as part number 9779072 and it was painted Pontiac Silver Blue.

1968 Firebird Intake Manifold Part Number

Carburetor	Part Number	Notes
	230-ci Six Cylinder	
1-bbl	9794231	
4-bbl	9794232	
	350-ci V-8	
2-bbl	9794233	
4-bbl	9794234	
	400-ci V-8	
4-bbl	9794234	

1969

Most Firebird intakes were painted Pontiac Silver-Blue to match the engine block the exceptions were those built with the Ram IV option package. This package added an aluminum intake. The casting date was in the same location that was used in 1968 and while it is important that this date be before the assembly date of your Firebird, it is equally important that this date be around the same casting date that the other engine components were cast (block, head) to be consider correct. Dates do not have to be exact; it was common to have from 1- day to 2 weeks apart from the casting date on the other engine components

Ram Air

1967

Ask any Firebird owner what one package would they like to own; and the answer will come back: Ram Air. The Ram Air went through many different phases. The unit in 1967 was known as Ram Air, was available on the Firebird 400 model only. It was available from the factory, even thought there were no sales codes for this option it was S.T.O.P for Pontiac, as there was speculation that the Ram Air option would be dropped in 1967 or made available only as a dealer installed option. The second way the Ram Air package was available was as a dealer

1969 Firebird Intake Manifold Part Numbers

Carburetor	Part Number	Notes
	230-ci Six Cylinder	
1-bbl	9794231	
4-bbl	9794232	
	350-ci V-8	
2-bbl	9794233	
4-bbl	9794234	
	400-ci V-8	
4-bbl	9794234	Except Ram Air IV
4-bbl	9799084	Ram Air IV

installed option package listed as part number 984887 it included a special panel and seal, with a black colored lid for the air cleaner, air filter element a closed-type breather tube and, a special Ram Air name plate for the hood and instruction. The instructions told how to install the panel and open up the sealed hood scoops on the Firebird. Cost of this package was around $100.00 and it looked a lot like the GTO Ram Air set up. No records are kept on how many of these dealer installed packages were installed. The only way to document this package is with a written work order to install the package.

The gloss black shroud and foam seal were sold together under part number 9789884, while they are similar in design, a unit from a GTO will not fit a Firebird. A seal part number 9789855 was used between the shroud and the air inlet baffle and a seal part number 9789883 was used between the inlet baffle and the hood. The cover was listed as part number 6424398, and should be painted 60-degree gloss black, as should the bottom shroud.

The same carburetor that was used on the standard 400 ci HO, was also used on the Ram Air package with manual transmission. A special carburetor was used with the Ram Air package when the automatic transmission was ordered. Listed as part number 7037273, this carburetor and the manual unit were used with and without A.I.R. emission packages.

For the first part of the year the Ram Air package used the same cylinder heads that were used on the 400-ci HO Firebird. Towards the end of the model year starting with engine serial number 546616 a new cylinder head part number 9783657 casting number 97 was used. Note that the casting number for this head was not in the usual center location, but next to the lower head bolt and behind the center exhaust port. The big thing that was changed about this head was all new valves and tougher valve springs. This head was used only in car with factory installed Ram Air. Those with the dealer installed package continued to use the standard 400-ci HO heads.

1968

The dealer installed Ram Air package was deleted and all Ram Air cars came from the factory only. The early style (up to March 15, 1968) used the common D-port design cylinder heads listed as part number 9792700, it can be identified by the casting "37". Those with Ram Air used special Rochester 4MV Quardajet carburetors. Part number 7028276 was used with manual transmissions and part number 7028277 was used with automatics.

On March 15, 1968 the Ram Air I package, as it is known now, was deleted and new Ram Air package took its place. Ram Air II was the forerunner to the legendary Ram Air IV packages in 1969. The Ram Air II used a special 4-bolt mains that was loaded with special forged pistons and high riser camshaft. Plus, the D-port heads were deleted and new round port heads part number 9794040-cast number 96. The block used special ID codes of WU for manual transmission and WY for automatic. Warning: Be extra careful when identifying this block as the WU code was used in 1969 -1972 on the common 350-ci block. When looking for this block, be sure that the casting date shows it is for 1968, but not cast after July 1968 which would make it for a 1969 model and be a 350-ci block. No matter the casting date also check the VIN stamp on the block. While you cannot document the model the engine came from, you can check the assembly plant (remember all 1968

Firebirds were built at the Lordstown plant and will have the code "U" in the third digit of the stamp) and more importantly, the model year. Be aware that the GTO (although they used different codes) also used this powerplant so a true Ram Air may have different factory codes. But only a block with the Lordstown code is considered correct for a Firebird.

Ram Air II also used Rochester Quardajet carburetors. However, research shows that special units were used. Those with manual transmission use ID number 7028273 and those with automatic ID number 7028270.

The air induction was a hybrid system, consisting of new parts and parts that were a carry over from 1967. While the Ram Air II package used different mechanical parts, the air induction system was the same all year long.

The shroud and seal part number 9792987 were restyled and more rounded in shape. The seals used the same part numbers as they used in 1967, but the hood inlet baffle was new and given part number 9793161.As before the shroud and the cover, which also used the same part number, should be painted 60-degrees gloss black.

1969

There were two different Ram Air Systems that were available in 1969 Firebirds. The basic Ram Air set up was the Ram Air III package. Taking a step backwards in performance the Ram Air III package used the D-port heads part number 9795043 which were the same units used on the standard Firebird 400. Like wise the Ram Air III package used the same cast iron intake manifold as the standard 400-ci V-8 did. As for carburetors the Ram Air III package used the same carburetors the Ram Air II package used in 1968 ½.

The hotter set up was the Ram Air IV package, this was an improvement over the 1968-1/2 Ram Air II package. It used the same block, and internal accessories. The cylinder head, part number 9799498, was restyled to use longer pushrods, anti-pump-up lifters and a 1.65 in. rocker arm package. Also an aluminum intake manifold with matching ports was also used, which gave the powerplant an all around better drivability to it. Special Quardajet carburetors with Part ID number 7029273 with manual transmission and 7029273 with automatic transmission, were used.

The 1969 Ram air inlet air cleaner assembly was completely restyled. The base (shroud) had a ununiform design with ears sticking out along the sides of the base. It was listed as part number 9798121 for use with Ram Air III and part number 9798113 with the Ram IV set up. These bases had a general similar design, the difference being that the crank case vent hoses attach. Those with the Ram Air IV used a preheater tube while the Ram Air II was routed directly into the right-hand valve cover. Other components also differed:

Ram Air Baffle and seal Part Numbers

Part	Ram Air III	Ram Air IV
Seal shroud to baffle	9798027	9797913
Air Filter Element	6485167	6485167
Baffle	9797838	9797839
	546284*	546285*
Seal Element	9797438	9797438
Seal Baffle to Hood	9798061	9798061
	546282*	546282*

The 1969 Trans Am used a special hood with air ducts located at the far front end of the hood, this required special ductwork part numbers 546012 right-hand and 546013 left-hand to direct the air from the hoodscoops to the carburetor. One of the complaints about the older style Ram Air packages was that during rainy or snowy weather rain would get into the engine compartment through the open hoodscoops. In 1969 a simple cable format gave protection during adverse weather. It consisted of a cable that was attached to the lower portion of the instrument panel on the driver's side. Important two styles of cables were used. First type was a bare cable while the second type used a black plastic coated cable. There are slight differences in the parts that mount these controls. For correctness and the proper operating procedure the proper parts must be used.

The cable was routed through the firewall and attached to an inlet valve part number 9798191. When the control was in the closed position the inlet valve would shut and close air from being directed down the carburetor, instead the engine used the under hood air. This was good in cold weather. When performance was needed in good weather the control could be opened and the valve would open and allow cool outside air into the carburetor.

Ram Air Control Parts

Part	1st type	2nd type
Control	N/A	546527
Bracket to Baffle	9794928	546517
Bushing	9797418	9797418
Cable inlet to valve shaft	546524	9797438
Connector inlet to air valve	546143	Not used
Connector	Not used	3923213

Another view of the odd shaped hood seal that was used with the 1969 Ram Air set up.

Exhaust Systems

1967-1969

All Firebird models used cast iron exhaust manifolds. They should have a fresh cast appearance; painting them with high temp cast iron gray paint can duplicate this. To get the best finish you should start with a clean surface. First clean the manifold with a stiff wire brush, wear gloves and eye protection when doing this procedure to avoid injuries. For manifolds that are very rusty, they may have to be bead blasted. Then using compressed air throughly blow clean, being sure to get into all the crevasses. Then apply a thin coat of the paint, allow to dry for 1-2 hours then apply second coat. Allow to dry for 24-hours. Bake the part at 400-degrees for 1-hour, unless you're looking for a divorce do not use the oven in the kitchen, it is best you have separate oven for the curing method. If this is not possible allow the heat of the engine to cure the paint. The finish may not have as much durability as the baking process but it will cure.

Galvanized reverse flow mufflers were used all three years. The 4-bbl six cylinders and the V-8s with single exhaust all used the same muffler, which was 25/14 in. length and 9-3/4 in. wide. They were listed as part number 9789774 all three years. The unit used on a 1-bbl six cylinder was listed as part number 9789773. The head pipe should have a natural finish to it, as should the tail pipe. Dual exhaust was standard on all six - cylinders ,2-bbl V-8's and the 1968-69 350-ci 4-bbl. All 400-ci V-8s came standard with dual exhaust. Dual exhaust was optional on the 350-ci 4-bbl V-8 in 1968-1969, it was standard in 1967. While the standard pipes are correct for the cars, the draw back to these pipes was that the system did not last. They would rust and rot. Even try to pick up an original exhaust system for a 20 plus year old car in the salvage yard? The unit will literally crumble in your hand. Great for you morale, makes you feel like Superman. Bad for a show car that is driven. A better solution is a stainless steel exhaust system. It will cost more but they last nearly forever. Plus most car shows do not take off points for this kind of system, if it has the same look and route that the original unit used.

1967 Firebird Exhaust Systems

Engine	Manifold	Head pipe	Tail Pipe	Notes
230-ci 1-bbl	545977	9777669-front 9789701-rear	9796870	
230-ci 4-bbl	9786239	9777678-front 9777668- middle 9789700- rear	9796870	
326-ci 2-bbl	9777755-L.H 9779325-R.H..	9777628- front 9789700-rear	9796870	
326-ci 4-bbl	9777755-L.H 9779325-R.H..	9777672- L.H. Front 9777670-R.H. Front 9789700- L.H. Rear 9797467-R.H. Rear	9789703-L.H. 9796869-R.H	
400-ci 4-bbl	9777755-L.H 9779325-R.H..	9777672- L.H Front. 9777670-R.H. Front 9789700- L.H. Rear 9797467-R.H. Rear	9789703-L.H. 9796869-R.H	Except Ram Air
400- Ram Air	9779495-L.H 9779493-R.H..	9781967- L.H. Front 9781965-R.H. Front 9789700- L.H. Rear 9797467-R.H. Rear	9789703-L.H. 9796869-R.H	Ram Air

1968 Firebird Exhaust Systems

Engine	Manifold	Head pipe	Tail Pipe	Notes
250-ci 1-bbl	545977	9777669-front 9789701-rear	9796870	
250-ci 4-bbl	9790443	9777716-front 9792647-middle 9789700- rear	9796870	
350-ci 2-bbl	9777755-L.H 9779325-R.H..	9777628- front 9789700-rear	9796870	
350-ci 4-bbl	9777755-L.H 9779325-R.H..	Single 9777628- front 9789700-rear Dual 9777672- L.H. Front 9777670-R.H. Front 9789700- L.H. Rear 9797467-R.H. Rear	9789703-L.H. 9796869-R.H	
400-ci 4-bbl	9777755-L.H 9779325-R.H..	9777672- L.H. Front 9777670-R.H. Front 9789700- L.H. Rear 9797467-R.H. Rear	9789703-L.H. 9796869-R.H	Except Ram Air
400- Ram Air	9779495-L.H 9791607-R.H..	9781967- L.H. Front 9781965-R.H. Front 9789700- L.H. Rear 9797467-R.H. Rear	9789703-L.H. 9796869-R.H	Ram Air

1969 Firebird Exhaust Systems

Engine	Manifold	Head pipe	Tail Pipe	Notes
250-ci 1-bbl	545991	9777669-front 9789701-rear	9796870	
250-ci 4-bbl	9790443	9777716-front 9792647-middle 9789700- rear	9796870	
350-ci 2-bbl	490144-L.H 490142-R.H..	9777628- front 9789700-rear	9796870	
350-ci 4-bbl	490144-L.H 490142-R.H..	Single 9777628- front 9789700-rear Dual 9777672- L.H. Front 9777670-R.H. Front 9789700- L.H. Rear 9797467-R.H. Rear	Single 9796870 Dual 9796797-L.H. 9796869-R.H	
400-ci 4-bbl	490144-L.H 490142-R.H..	9796600- L.H. Front 9797132 R.H. Front 9789700- L.H. Rear 9797467-R.H. Rear	9796797-L.H. 9796869-R.H	Except Ram Air
400- Ram Air III	9779495-L.H 9797073R.H..	9781967- L.H. Front 9781965-R.H. Front 9789700- L.H. Rear 9797467-R.H. Rear	9796797-L.H. 9796869-R.H	Ram Air III
Ram Air IV	9794038-L.H. 9797074-R.H.	9781967- L.H. Front 9781965-R.H. Front 9789700- L.H. Rear 9797467-R.H. Rear	9796797-L.H. 9796869-R.H	Ram Air IV

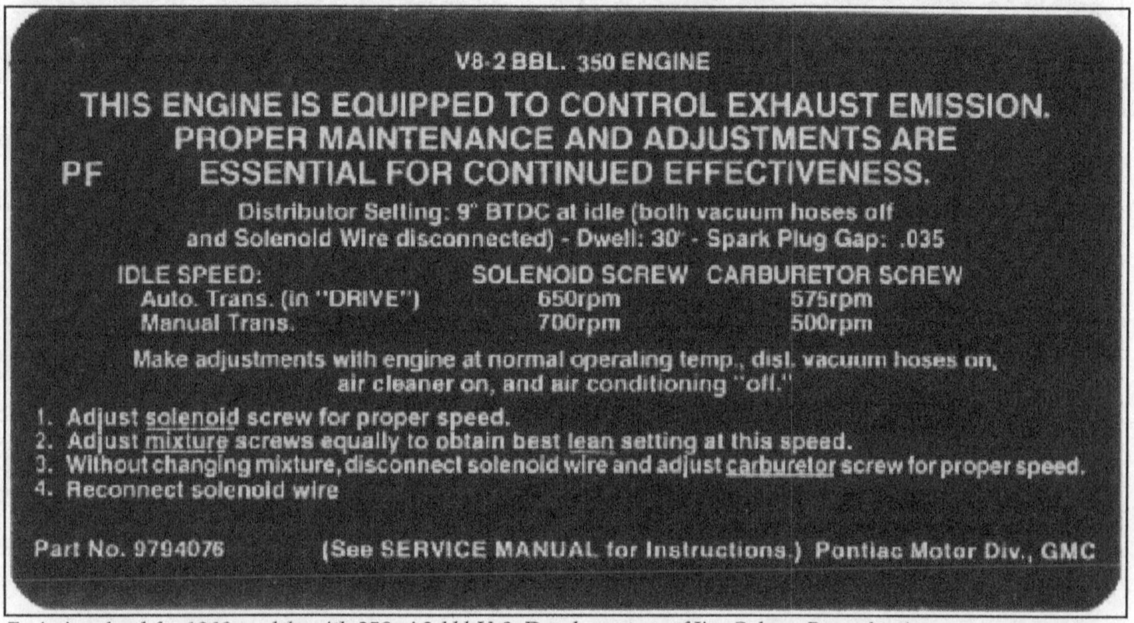

Emission decal for 1968 models with 350-ci 2-bbl V-8. Decal courtesy of Jim Osborn Reproductions.

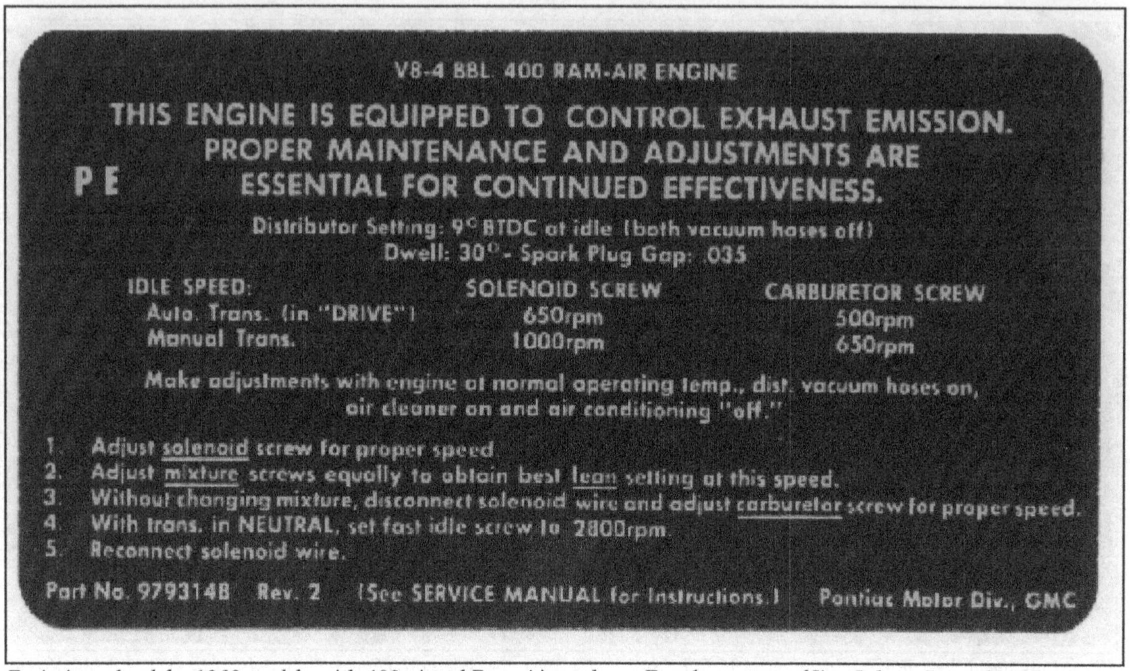

Emissions decal for 1968 models with 400-ci and Ram Air package. Decal courtesy of Jim Osborn Reproductions.

Emissions decal for 1968 Firebirds with a 350-ci or 400-ci with 4-bbl carburetor, except those with Ram Air. Decal courtesy of Jim Osborn Reproductions.

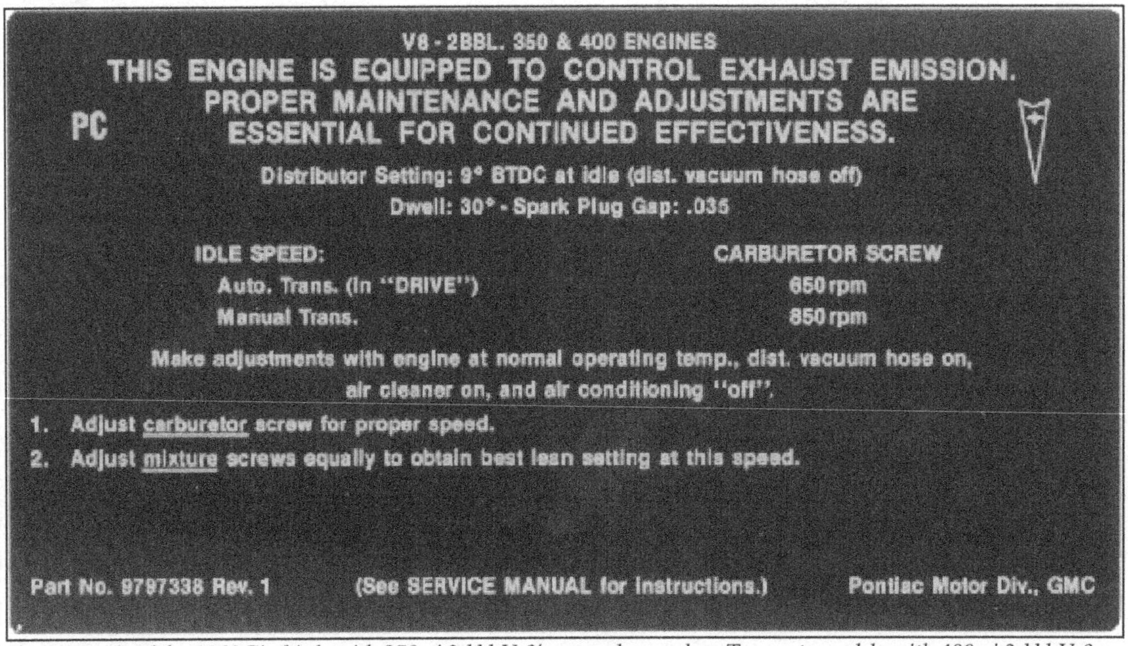

Emissions decal for 1969 Firebirds with 350-ci 2-bbl V-8's, was also used on Tempests models with 400-ci 2-bbl V-8.s Decal courtesy of Jim Osborn Reproductions.

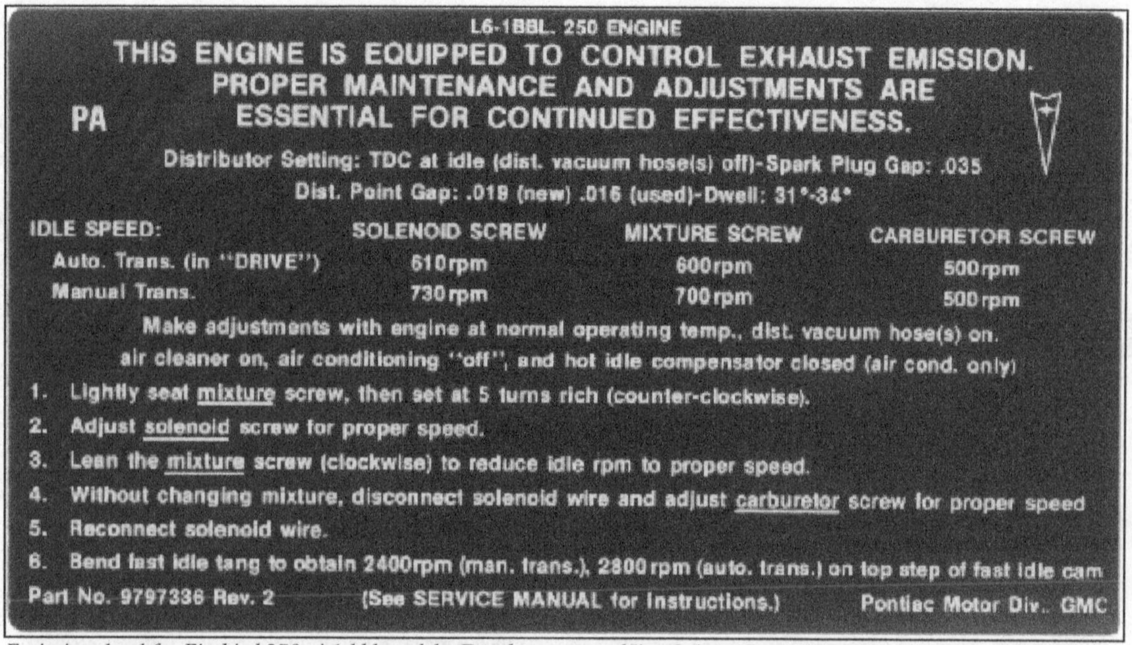
L6-1BBL. 250 ENGINE

THIS ENGINE IS EQUIPPED TO CONTROL EXHAUST EMISSION.
PROPER MAINTENANCE AND ADJUSTMENTS ARE
PA ESSENTIAL FOR CONTINUED EFFECTIVENESS.

Distributor Setting: TDC at idle (dist. vacuum hose(s) off)-Spark Plug Gap: .035
Dist. Point Gap: .019 (new) .016 (used)-Dwell: 31°-34°

IDLE SPEED:	SOLENOID SCREW	MIXTURE SCREW	CARBURETOR SCREW
Auto. Trans. (in "DRIVE")	610rpm	600rpm	500rpm
Manual Trans.	730rpm	700rpm	500rpm

Make adjustments with engine at normal operating temp., dist. vacuum hose(s) on, air cleaner on, air conditioning "off", and hot idle compensator closed (air cond. only)

1. Lightly seat mixture screw, then set at 5 turns rich (counter-clockwise).
2. Adjust solenoid screw for proper speed.
3. Lean the mixture screw (clockwise) to reduce idle rpm to proper speed.
4. Without changing mixture, disconnect solenoid wire and adjust carburetor screw for proper speed
5. Reconnect solenoid wire.
6. Bend fast idle tang to obtain 2400rpm (man. trans.), 2800rpm (auto. trans.) on top step of fast idle cam

Part No. 9797336 Rev. 2 (See SERVICE MANUAL for instructions.) Pontiac Motor Div., GMC

Emission decal for Firebird 250-ci 1-bbl models. Decal courtesy of Jim Osborn Reproductions.

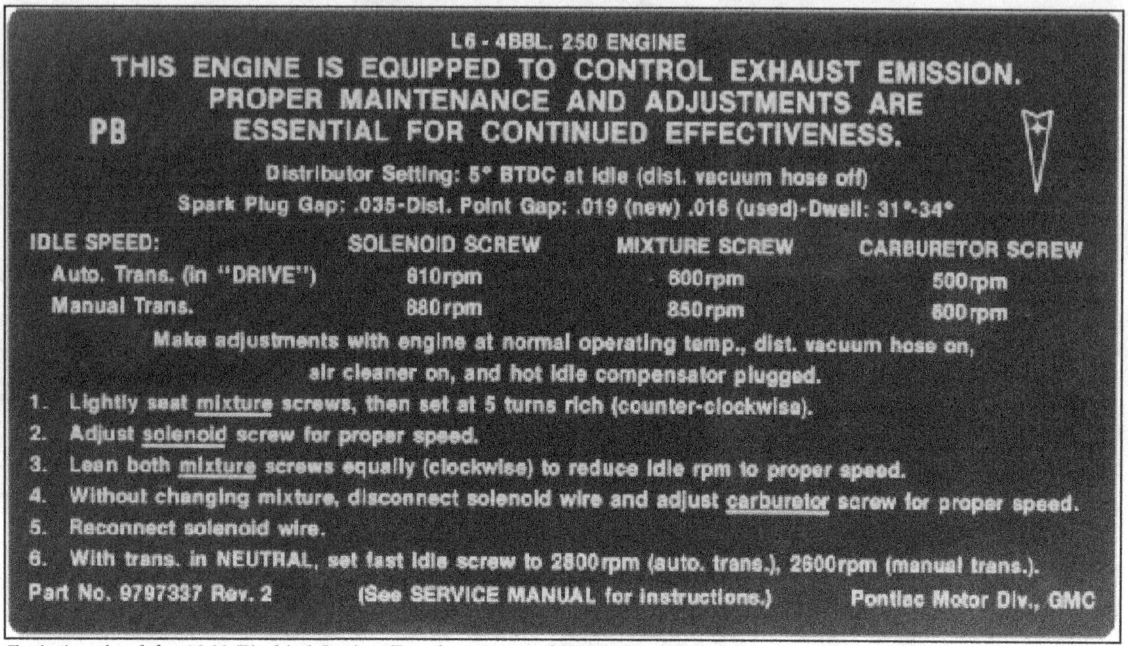
L6 - 4BBL. 250 ENGINE

THIS ENGINE IS EQUIPPED TO CONTROL EXHAUST EMISSION.
PROPER MAINTENANCE AND ADJUSTMENTS ARE
PB ESSENTIAL FOR CONTINUED EFFECTIVENESS.

Distributor Setting: 5° BTDC at idle (dist. vacuum hose off)
Spark Plug Gap: .035-Dist. Point Gap: .019 (new) .016 (used)-Dwell: 31°-34°

IDLE SPEED:	SOLENOID SCREW	MIXTURE SCREW	CARBURETOR SCREW
Auto. Trans. (in "DRIVE")	610rpm	600rpm	500rpm
Manual Trans.	880rpm	850rpm	500rpm

Make adjustments with engine at normal operating temp., dist. vacuum hose on, air cleaner on, and hot idle compensator plugged.

1. Lightly seat mixture screws, then set at 5 turns rich (counter-clockwise).
2. Adjust solenoid screw for proper speed.
3. Lean both mixture screws equally (clockwise) to reduce idle rpm to proper speed.
4. Without changing mixture, disconnect solenoid wire and adjust carburetor screw for proper speed.
5. Reconnect solenoid wire.
6. With trans. in NEUTRAL, set fast idle screw to 2800rpm (auto. trans.), 2600rpm (manual trans.).

Part No. 9797337 Rev. 2 (See SERVICE MANUAL for instructions.) Pontiac Motor Div., GMC

Emission decal for 1969 Firebird Sprint. Decal courtesy of Jim Osborn Reproductions.

1967 A.I.R. emission decal (Florescent orange background.) Decal courtesy of Jim Osborn Reproductions.

An example of the location of an emissions decal, shown is on the left-hand side of the filler panel, but they have been found on the right-hand side also.

An example of the location of an air cleaner service decal location.

Cooling Systems
RADIATOR AND FAN

A cross flow radiator core was used in all Firebirds. A cross flow core flows coolant horizontally and the tanks are on the sides, instead of above and below as in the conventional down-flow type core. The cross-flow core provides for better cooling properties. Firebirds with automatic transmissions have coolers built into the right tank with the inlet and outlet fitting for the circulation of the transmission field. Firebirds with air conditioning, heavy cooling and those models with the A.I.R. emission packages require special heavy-duty cores. The core should be painted flat black. A natural appearing 15psi pressure cap was used on all cores.

A four-blade fan (part number 9798940) was standard on all models except for a few exceptions. This fan was 19-in. diameter and had a 2-5/16 in. pitch. It used an aluminum spacer, which should have a natural appearance. With the six-cylinder a 1.00-in. spacer part number 3927792 was used. For the V-8 a 2-1/2 in. spacer part number 535253 was used. The exception to the standard fan the Firebird 400 models without the Ram Air package or when air conditioning or heavy-duty cooling was ordered.

When air conditioning was ordered then a seven-blade fan was part number 9788982 for six-cylinder was used. This fan was 18 in. diameter. This unit, used on V-8s, was listed as part number 9777012, this 19-1/2 in diameter fan was the same unit used on the Firebird 400, without Ram Air. A fan clutch assembly drove it. The clutch assembly should have a cadmium gold finish and was marked with the letters "PH". California cars in 1967 with air conditioning used the same fan but the clutch assembly was listed as part number 9790972 and marked "PG".

The 326-ci V-8 was available with a five-blade heavy-duty fan listed as part number 9779005, this 19-in. diameter fan had a 2-1/4 in. pitch. It used the same spacer as the standard 4-blade fan did.

Firebird with the Ram Air package used the same 4-blade fan that was used on the base Firebirds. Except the 1969 Ram Air III package which used the standard 7-blade fan.

A 60 degree gloss black fan shroud part number 9788883 was standard on all V-8 Firebirds. A shroud was not used on six-cylinder Firebirds.

Radiator hoses were molded and the six cylinders and V-8 models used two different hoses. Both the inlet and outlet hoses have a inside diameter of 1-3/4 in.

1967-1969 Radiator Hoses

Engine	Inlet	Outlet
Six Cylinder	9788995-1967-68 9798648-1969	9788996
V-8	9788916	9788916

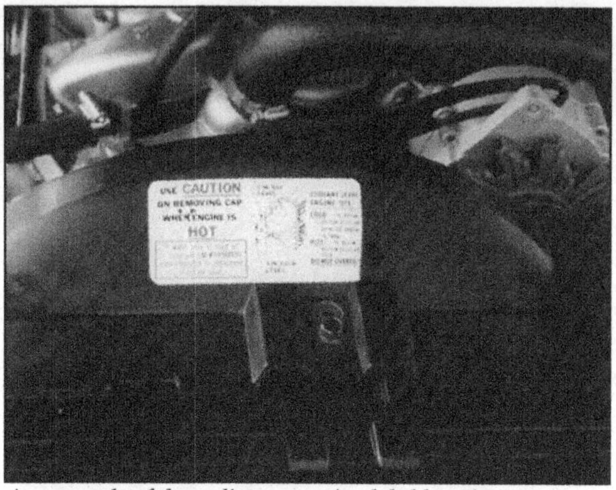

An example of the radiator warning label location, on the fan shroud. When no shroud is used then it is place on the radiator support wall.

1967 Radiator Core Identification Codes

Transmission	Standard		Heavy-Duty Cooling		Air Conditioning		Notes
			230-ci Six-Cylinder				
Manual	1-bbl	4-bbl	1-bbl	4-bbl	1-bbl	4-bbl	
	UT	UG	UG	US	US	US	
Automatic	UT	UT	UG	UG	US	US	
			326-ci V-8				
Manual	2-bbl	4-bbl	2-bbl	4-bbl	2-bbl	4-bbl	
	UU	UU	UM	UM	UQ	UQ	
Automatic	UV	UW	UW	UX	UP	UP	*-With A.I.R
	UW*		UX*				
			400-ci 4-bbl V-8				
Manual	UQ		Std.		UQ		
Automatic	UP		Std.		UP		

Hose Clamps

Standard hose clamps were the tower style clamps. These clamps were quickly discarded and since GM did not offer them as a replacement, were usually replaced with worm gear type clamps, as they had better holding power. Big points can be had at a car show when the proper style tower clamps are used. In addition the original Wittek built clamps were date coded, according to the quarter of the production year that your Firebird was built. A typical date code will be the quarter followed by the year of manufacture. It is important to remember that the first quarter may not agree with the model year. For example the date code 3Q6 would be the third quarter of the 1966-year and be for an early built 1967 model. Also while it is listed in quarters most parts were stamped only as early (third quarter) or late (first quarter) production runs. This corresponds with model introduction and usual major changes in production. First quarter is January to March ; 2-nd quarter April to June ; 3rd quarter is July to September ; 4th quarter is October to December. However usually only the 1st and 3rd quarter date codes are used. As production for the clamps began before and after release of new models, so most clamps are coded either 1Q or 3Q.

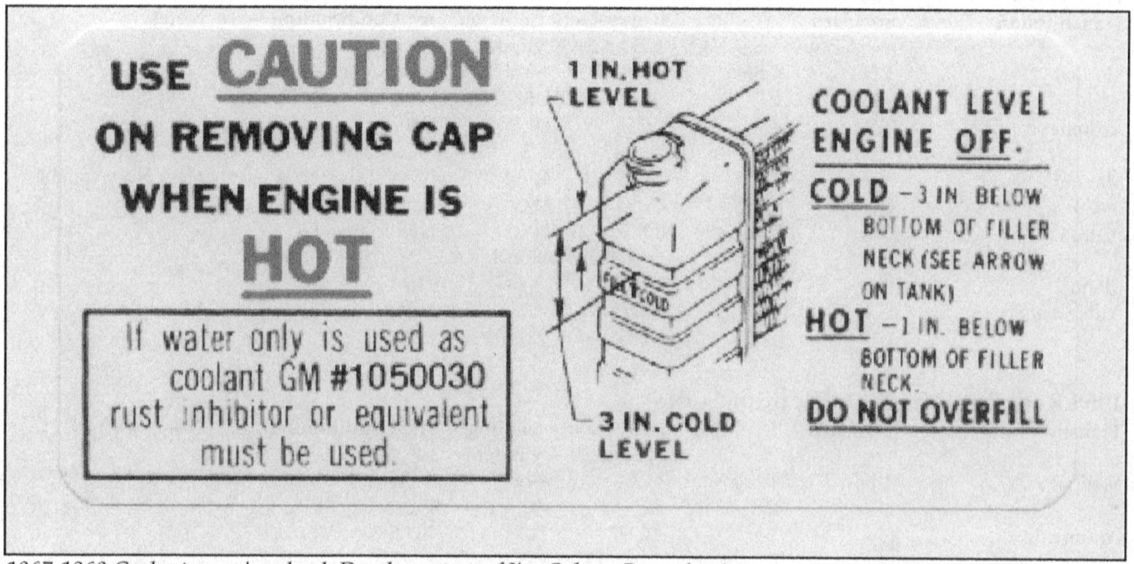

1967-1968 Coolant warning decal. Decal courtesy of Jim Osborn Reproductions.

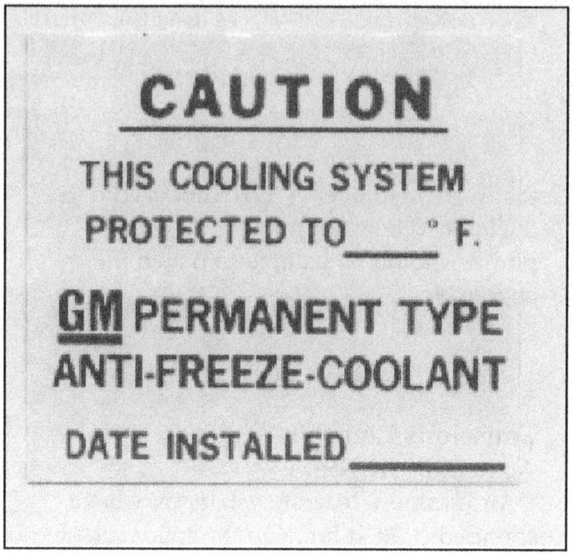

Anti-freeze coolant label. The blanks were filled in by hand at the factory. Decal courtesy of Jim Osborn Reproductions.

1968 Radiator Core Identification Codes

Transmission	Standard		Heavy-Duty Cooling		Air Conditioning		Notes
			250-ci Six-Cylinder				
Manual	1-bbl	4-bbl	1-bbl	4-bbl	1-bbl	4-bbl	
	UE	UE	US	US	US	US	
Automatic	ZA	ZA	UF	UF	UR	UR	
			350-ci V-8				
Manual	2-bbl	4-bbl	2-bbl	4-bbl	N/A	N/A	N/A Not available
	UU	UU	UX	UX			
Automatic	UW	UW	UX	UX	UP	UP	
			400-ci 4-bbl V-8				
Manual	UQ		UQ		UQ		
Automatic	UX		UX		UX		* Ram Air
	UP*		UP*		UP*		

1969 Radiator Core Identification Codes

Transmission	Standard		Heavy-Duty Cooling		Air Conditioning		Notes
			250-ci Six-Cylinder				
Manual	1-bbl	4-bbl	1-bbl	4-bbl	1-bbl	4-bbl	
	UE	UE	ZC	ZC	ZC	ZC	
Automatic	ZA	ZA	ZC	ZC	N/A	N/A	
			350-ci V-8				
Manual	2-bbl	4-bbl	2-bbl	4-bbl	2-bbl	4-bbl	N/A Not available
	UU	UM	UM	UQ	UQ	UQ	
Automatic	UW	UX	UX	UP	UP	UP	
			400-ci 4-bbl V-8				
Manual	UM#		UQ#		UQ#		# Base/ Ram Air III
	UQ*						
Automatic	UW#		UP#		UP#		*Ram Air IV
	UP						

Water Pumps

There are four different water pumps used on the 1967-1969 Firebird line up. The six cylinder 230-ci in 1967 and the 250-ci in 1968 used part number 9784280 and can be identified by casting number 9782041. The V-8 used part number 9793130, and has eight mounting holes.

Water pumps were restyled in 1969. The six cylinders used part number 9797609, this unit will not fit earlier models. The 1969 V-8's used part number 9797581, which has 11-mounting holes and is commonly referred to as a short snout pump as the distance from the blocking mounting surface and the fan hub surface is 4.00 inches. Close to the end of production year this pump was phased out and replaced with part number 478126. This later unit is known as a long snout as it has a measuring distance of 4-1/2 in. from block surface to surface of fan hub. There is no set date as to when this switch occurred, but it believed that all 1969 style Firebirds that were held over as 1970 models did use the later style water pump. All water pumps should be painted to match the engine block.

Emissions Controls

A.I.R. EMISSIONS

Air Injection Reactor system, may have sounded right at home in the space age of the 1960's as something about a starship, but this package, even though it is well known was very limited in use. It was sold only on cars that were to be registered in California. Cars sold in any of the other 49 states did not come with the A.I.R. emissions.

The A.I.R. emission system consisted of a belt-driven air injection pump, specially design cylinder heads that contained air passages to the rear of each exhaust valve. Also special carburetor and air cleaner with provisions for the emission hoses, also a special distributor that altered the advance was also required in most cases.

A total of three different pumps were used with this package. Those with six-cylinder engines, the type of steering affected their usage. Those with standard steering used part number 7800752, while those with power steering used part number 7800751. All V-8 powered Firebirds in California used part number 7900307 regardless of the type steering ordered. The pump was mounted to engine with a 60-degree gloss black bracket research show some were painted engine blue. The bracket part number was dependent on the pump. An adjustment, natural aluminum appearing brace was also used on the pump.

1967 A.I.R. Pump Braces And Brackets

Engine	Steering	Bracket	Brace
230-ci	Std.	9782687	9783140
	P/S	9782687	9783140
326/400	All	9786814	9786813

The pulley used on the A.I. R. pump was painted 60-degree gloss black. A single or twin groove pulley was used; the part number depended on engine type and certain options on the car. A belt pulled the pulley.

1967 A.I.R. Pump Pulley and Belts

Engine	Steering	Part Number	Grooves
230-ci	Std.	9778560*	1
	Std.	9778561**	2
	P/S	9782690	1
326/400	All	9785478*	1
	All	9782664**	1

*- W/o a/c **-with a/c

1967 A.I.R. Pump Pulley Belts

Engine	Steering	Part Number	Dimensions
230-ci	Std.	9789230*	1/2x52 #
	Std.	9779516**	3/8x27-5/8
	P/S	See Power steering belts	
326/400	Std.	9786812	3/8x69
	P/S	382645	3/8x61

*- W/o a/c **-with a/c #-Double warp

The diverter valve part number 9788557 should have a natural appearance and was used on all Firebird models with the A.I.R. emissions. While the valve was the same with six-cylinders and V-8s, the mounting bracket was not. The following brackets were used to mount the diverter valve: 9777813-six cylinder with standard type steering; 9777814 six cylinder with power steering, and 9788551 for all V-8s.

Inlet hoses were broken down by engine choice. The outlet hose that ran from the outlet valve to the diverter valve was broken down by engine type and type of steering option in six cylinders. Tower clamps were used to hold these hoses in place. From the carburetor to the bleed valve (on V-8 Firebirds) a 25-in long hose part number 9781377 was used.

1967 A.I.R. Inlet and Outlet Hoses

Engine	Steering	Part Number	Notes
230-ci	Std.	9788117	Inlet
	P/S	9777816	Inlet
	Std.	9788119	Outlet
	P/S	9777815	Outlet
326-ci	All	9783544	Inlet (2-bbl)
	All	9788552	Outlet (2bbl)
	All	9783638	Inlet (4-bbl)
	All	9788552	Outlet (2bbl)
400-ci	All	9778593	Inlet (4-bbl)
	All	9788552	Outlet (2bbl)

POSITIVE CRANKCASE VENTILATION

This emission control system came in two forms, open and closed. All California bound Firebirds incorporated the closed type. The open type was available only on certain Firebirds that were not registered in California. For 1968 and 1969 the closed system is standard on all Firebird models.

The open type is very simple. It used a hose to connect to the air cleaner and directed crankcase vapors into the carburetor where they are burned again. A breather cap controls the emission flow. The six-cylinders used part number 6419440, while the V-8's used part number 1552232, except the Firebird 400, which used a chrome-plated breather cap listed as part number 6410204.

In the closed system the engine breathes filtered air through a tube running from the air cleaner back to the crankcase, crankcase vapors were then fed up through another tube to the carburetor, or air cleaner to be burned again in the carburetor. This required a fully enclosed breather cap. The following caps were used; 6420743- six cylinders and 850843 (type FC-2) with V-8. These caps were painted, while those used on the Firebird 400 part number 6410204 were chrome plated. A PCV valve part numbers 6423581 (type CV-735C) for six-cylinders and 6421972 (type CV679C) was used on the V-8 models.

CONTROLLED COMBUSTION SYSTEM

Not used on any Firebird in 1967, the C.C.S. first appeared as standard affair on all 1968 Firebirds, and continued into 1969. The system consisted of a sensor part number 6484242 that sensed when engine compartment temperatures were below 86-degrees and would keep the diaphragms, that were located on the snorkel (or the baffle in 1969 Ram Air systems) to remain closed, and thus the engine would draw on under hood warmed air resulting in quicker warm ups. Not much needed in southern California, but a true welcome in the cold midst of the winter in the midwest. And you though all emission controls were bad.

When the temperature reached about 100 degrees in the engine compartment the sensor would allow the diaphragms to open and draw in fresh air. Vacuum hoses were routed from the sensor to the diaphragms. The following diaphragms were used:

1968-69 Air Cleaner Diaphragms

Engine	Carb	Part Number
250-ci	All	6484245
350-ci	2-bbl	6484245
350-ci	4-bbl	6484247
400-ci	4-bbl (except Ram Air)	6484246
400-ci	Ram Air	6485088-LH 6485087-RH

Chapter 8

Body and Sheet Metal

Exterior Finish

There is a big difference in the quality of paint then (1967-1969) and now. Want to see what the original paint looks like? Take a look at old TV shows. Like Roger's Firebird convertible in *I Dream of Jeannie.* Good reason to make an excuse on watching TV "*Hey I am doing research!*" You will notice that while it shines, is not as deep or as flawless as most show cars are today. The main reason for this is the care that is placed into painting a classic car today, while when it was new the paint was quickly applied, and even runs can be seen. But also the quality of paint has much improved. Now I am not suggesting that you haphazardly apply your paint to your car to duplicate the factory look. That is one area that would not get you any points at a car show; in fact it would get you a deduction in points. The purpose of a restoration is not to look just look like it came from the factory, but to look like it came from the factory, if they were building only a classic, and took care and time to hand assemble each part. Which would mean a flawless paint job. So when applying your paint take extra care. A quality paint job can make the difference between the best of show and going home empty handed.

For all three years Firebird were painted with acrylic paint. The original color is coded on the body tag that is attached to the firewall (see chapter 1 for more details on locating and decoding this tag.)

1967 Exterior Color Codes

Code	Color Name	Du Pont	Ditzler	Rinshed-Mason
A	Starlight Black	88	9300	A-946
C	Cameo White	4024-L	8259	A-1199
D	Montreux Blue	4815-L	13349	A-1899
E	Fathom Blue	4817-L	13346	A-1900
F	Tyrol Blue	4850-L	13364	A-1920
G	Signet Gold	4825-L	22818	A-1919
H	Linden Green	4816-L	43651	A-1901
K	Gulf Turquoise	4818-L	43661	A-1903
L	Mariner Turquoise	4824-L	43659	A-1904G
M	Plum Mist	4832-L	50717	A-1905
N	Burgundy	4624-L	50700	A-1711M
P	Silver Glaze	4820-L	32603	A-1906
R	Regimental Red	4822-LH	71583	A-1907R
S	Champagne	4826-L	22813	A-1908
T	Montego Cream	4819-L	81578	A-1909

1968 Exterior Color Codes

Code	Color Name	Du Pont	Ditzler	Rinshed-Mason
A	Starlight Black	88	9300	A-946
C	Cameo White	4024-L	8259	A-1199
D	Alpine Blue	4892-L	13512	A – 1985
E	Agenda Blue	4899-L	13513	A-1992
F	Nordic Blue	4901-L	13514	A-1994
G	April Gold	4896-L	22942	A-1988
I	Autumn Bronze	4904-L	60517	A-1996F
K	Meridian Turquoise	4900-L	13517	A-1993
L	Aleutian Blue	4893-L	13516	A-1986
N	Flambeau Burgundy	4915-LH	50775	A-1999F
P	Silvermist Green	4897-L	43774	A-1989
Q	Verdoro Green	4947-LH	43745	A-1936
R	Solar Red	4948-LH	71634	A-1997F
T	Primavera Beige	4895-L	81617	A-1987
V	Nightshade Green	4898-L	43773	A-1990
Y	Mayfair Maize	4620-L	81500	A-1715

1969 Exterior Color Codes

Code	Color Name	Du Pont	Ditzler	Rinshed-Mason
10	Starlight Black	88	9300	A-946
40	Mayfair Maize	5036-L	81500	A-1715
50	Cameo White	5033-L	2058	A-2080
51	Liberty Blue	5016-L	2075	A-2098
52	Matador Red	5009-LH	2076	A-2009-R
53	Warwick Blue	5015-L	2077	A-2100
55	Crystal Turquoise	5014-L	2078	A-2101
57	Midnight Green	5013-L	2079	A-2102
59	Limelight Green	5012-L	2080	A-2103
61	Espresso Brown	5011-L	2081	A-2104
63	Champagne	5064-L	22813	A-2105
65	Antique Gold	5010-L	2082	A-2106D
67	Burgundy	5063-LH	50700	A-2107M
69	Palladium Silver	5032-L	2059	A-2108
72	Carousel Red	5021-LM	2084	A-2111R
73	Verdoro Green	5066-LH	2095	A-2110
76	Goldenrod Yellow	5026-LH	2094	A-2119
86	Claret Red	5074-L	71763	169V51R
87	Windward Blue	4962-L	13759	A-2123
88	Nocturne Blue	5075-L	13841	169V29
89	Castilian Bronze	5076-L	23215	169V81

Front End Sheet Metal

1967

Standard hood for all models except the Firebird 400 was listed as part number 9788846. The Firebird 400 models used a special hood that featured twin hood scoops. This hood was listed as part number 9789418. On those models without Ram Air ornaments part numbers 9777957 right-hand and 9777958- left-hand were placed in the ends of the hoodscoops, making them non-functional. These ornaments should be painted the same color as the hood. When Ram Air was ordered, special ornaments that were open were used. Engine call outs were placed along the side of the center dome or the sides of the hood scoops on all models.

The V-8 callouts were chrome-plated and inlaid in red. The callouts used on the six-cylinders read "OVERHEAD CAM - 3.8 Liter" and were unique by the fact that they were color-keyed to the exterior color of the car. Most reproduction parts are black, and this has to be changed for correctness.

1967 Hood Callouts-V-8

Engine	Callout	R.H.	L.H.
326-ci	326	9777847	9777848
400-ci	400	9789482	9789483

1967 Hood Callouts-Six-Cylinder

Exterior Color	R.H.	L.H.
Starlight Black	9789380	9789381
Cameo Ivory	9789382	9789383
Montreux Blue	9789384	9789385
Fathom Blue	9789386	9789387
Tyrol Blue	9789388	9789389
Signet Gold	9789390	9789391
Linden Green	9789392	9789393
Gulf Turquoise	9789394	9789395
Mariner Turquoise	9789396	9789397
Plum Mist	9789398	9789399
Burgundy	9789400	9789401
Silver Glaze	9789402	9789403
Regimental Red	9789404	9789405
Champagne	9789406	9789407
Montego Cream	9789408	9789409
Verdoro Green	9789410	9789411
Coronado Gold	9789412	9789413
Mayfair Maize	9789414	9789415

Both hoods used the same set of hinges listed as part number 3910667 left-hand and 3910668 right-hand and hood springs part number 3926800. The hinges and springs were black cadmium some had a natural appearance. A hood pad part number 9788847 was standard, it was held in place with flat black retainer clips.

The hood latch part number 9712100 was attached to the vertical center brace of the radiator support wall. The latch assembly should have a natural appearance as should the latch plate on the hood. Hood pop up spring part number 548752 should have a flat black appearance. A black rubber seal part number 9789927 was used on the hood cowl ledge.

1967 hood callout for 230-ci six cylinder, note the inlaid body color.

1967 326-ci callouts on the hood

Front fenders were listed as part numbers 9793816- right-hand and 9793817-left-hand and were used on all models and body styles. Lower extensions were listed as part numbers 3891628 right-hand and 3891627 left-hand. The Firebird nameplate part number 9789532 was inlaid in white and was placed on the forward edge of the front fenders of all models. It was held on with three nuts. Placed in front of the Firebird name was the Firebird medallion. This red inlaid bird was listed as part number 9789589. The tips of the wings were inlaid in black, as was the body, the tail feathers were inlaid in red, it was held on with natural appearing retainers. The inner fender wells should be painted 60-degrees gloss black. The front valance panel was listed as part number 9789297. Bright plated wheel trim moldings were listed as part numbers 9785880 right-hand and 9785881-left-hand. It was held in place with seven Phillips head screws.

1968

Very few changes occurred in 1968. The standard hood used the same part number as it used in 1967. However, the hood used on the Firebird 400 model was listed as part number 9793429. This hood still featured twin hood scoops. On those models without Ram Air ornaments part numbers 9777957 right-hand and 9777958- left-hand were placed in the ends of the hoodscoops, making them non-functional. These ornaments should be painted the same color as the hood. When Ram Air was ordered, special ornaments that were open were used. These were listed as part number 9794283- right-hand and 9794284- left-hand . Engine call outs were placed along the side of the center dome or the sides of the hood scoops on all models.

The V-8 callouts were chrome-plated and inlaid in red. The callouts used on the six-cylinders read "4.1 Liter OVERHEAD CAM" and were unique by the fact that they were color-keyed to the exterior color of the car. Most reproduction parts are black, and this has to be changed for correctness.

1968 Hood Callouts-V-8

Engine	Callout	R.H.	L.H.
350--ci	350	9792691	9792692
400-ci	400	9789482	9789483

1968 Hood Callouts-Six-Cylinder

Exterior Color	R.H.	L.H.
Starlight Black	9792381	9792381
Cameo Ivory	9792383	9792384
Red	9792385	9792386
Maize	9792387	9792388
Nightshade Green	9792389	9792390
Springtime Green	9792391	9792392
Burgundy	9792393	9792394
Aegean Blue	9792395	9792396
Alpine Blue	9792397	9792398
Meridan Turquoise	9792399	9792400
Gold	9792401	9792402
Beige	9792403	9792404
Nordic Blue	9792405	9792406
Aleutian Blue	9792407	9792408
Verdoro Green	9792409	9792410
Bronze	9792411	9792412

Both hoods used the same set of hinges and springs that were used the year before. The hoodpad was also the same.

The hood latch part number 9715566 was restyled and would fit not fit earlier models. Hood pop up spring part number 9790484 was also restyled.

Front fenders used the same exact part numbers as they did in 1967. Also the same was the Firebird name and medallion. All refinishing was also the same. The front valance panel was listed as part number 9789297. Also the same were the wheel trim moldings.

1969

The standard hood used the same part number as it used in 1967 and 1968. two different hoods were used on the Firebird 400, the standard hood with non-functional hood scoops was listed as part number 9793430 while that with the cable controlled Ram Air used part number 9797763. The standard 400 hood used fixed ornaments listed as part number 9777957 right-hand and 9777958 right-hand, these ornaments should be painted same color as the car. Those used on the cable operated Ram Air hood part numbers 9794283 right-hand and 9794284 left-hand should be painted black. However the hood used on the Firebird 400 model was listed as part number 9793429. This hood still featured twin hood scoops. On those models without Ram Air ornaments part numbers 99794283 right-hand and 9784284- left-hand were painted flat black.

Trans Am model used a special hood with twin functional hoodscoops molded in at the forward edge of the hood. This hood was listed as part number 546014. This hood used special ornaments in the scoops listed as part numbers 546257-right-hand and 546328- left-hand. Hinges used the same part number as they did the two previous years. All except the Trans Am and Firebird 400 models used the same hood spring that was used on the previous years. Trans Am and Firebird 400 used a special low rated spring listed as part number 546325. No hood pad was used on the Trans Am model, all other models used the same pad that was used the years before.

Engine callouts for the V-8 models were again placed along the side of the center dome or the sides of the hood scoop, except on Trans Am or when the Ram Air package was ordered. When Ram Air was ordered then a decal was used instead. The decals came in either black, the most common, blue or red, with the latter of these the less common. The decals that were used with the Ram Air II package read "RAM AIR" it is incorrect for them to read RAM AIR II. The decals that were used on the Ram IV system did read Ram Air IV.

Callouts on the six cylinder cars were the same visually as those in 1968. And once again the inner portions of the nameplate were color-keyed to the exterior color. In fact those hold over colors from the previous year used the same part numbers.

1968 350-ci hood callouts inlaid in red.

1969 Hood Callouts-Six-Cylinder

Exterior Color	R.H.	L.H.
Yellow	9789132	9789133
Medium Blue	9789134	9789135
Starlight Black	9792381	9792381
Maize	9792387	9792388
Verdoro Green	9792409	9792410
White	9794887	9794888
Maroon	9798379	9798380
Champagne	9798381	9798382
Light Red	9798383	9798384
Silver	9798385	9798386
Dark Red	9798387	9798388
Gold	9798389	9798390
Brown	9798391	9798392
Lime	9798393	9798394
Dark Green	9798394	9798395
Turquoise	9798397	9798398
Light Blue	9798399	9798400
Dark Blue	9798401	9798402

Ram Air Hoodscoop Decals

Color	Ram Air III	Ram Air IV
Black	9794721	9789041
Red	9794722	9789042
White	9794723	9789043

Callout on the hood for the powerful 400-ci V-8, were placed along the sides of hoodscoops.

Front fenders were completely restyled, and made longer and more ranked and will not fit earlier models. The front fenders supported simulated air vents behind the front wheel openings. Trans Am models required special front fenders that featured functional air vents. Special air extractors part numbers 546252 right-hand and 546252 left-hand were placed over the vents on the fenders. Bright plated wheel well opening trim, listed as part numbers 9794773 right-hand and 9794774 left-hand were held in place with Phillips head screws. The trim molding was not used on Trans Am models.

1967-1968 Firebird nameplate, note the proper placement.

1969 Front Fender Part Numbers.

Model	Right-hand	Left-hand
Except T/A	9798212	9798212
Trans Am	546680	546681

The medallion on the front fender was deleted, and the Firebird nameplate was given part number 9796319, it was used on all models except the Trans Am. On the Trans Am the Firebird nameplate was removed and a blue color decal that read "TRANS AM" was used. Only the blue color decals were used. There are rumors that white was used on blue colored cars, but there is no evidence that these cars existed.

1967-1968 Firebird medallion

1969 Firebirds, except Trans Am, used these inserts. Upper moldings were listed as part numbers 9796361 right-hand 9796361 left-hand. Lower moldings were listed as part numbers 9796363 right-hand and 9796364 left –hand.

1969 Firebird fender nameplate.

1969 Trans Am .

Grille and Front Bumper

1967

A two-part grille assembly was used on all models. Two different grilles were used, all but the Firebird 400 models used part numbers 9788832 right-hand and 9788833 left-hand. The Firebird 400 grille was blacked out and listed as part number 9789373 right-hand and 9789374 left-hand.

The Pontiac nameplate part number 9788859 was used on the left-hand side grille insert of all models. The grille was housed in a twin loop chrome bumper. Two different bumpers were used. All but the Firebird 400's used part number 9788848 while those with the 400-ci power plant used part number 9789777, which used a special medallion listed as part number 9748667 in the center of the bumper. Note that if a shortage occurred of this bumper it is possible that a Firebird 400 may have been built with the standard bumper.

The hood latch and pop up spring should have a natural finish.

The Pontiac nameplate was used on the left-hand grille of all 1967 Firebirds.

Models with a 400-ci featured a arrow head medallion (the Pontiac logo) in the center of the front bumper

1969 front bumper medallion was used only on cars with 400-ci V-8.

1968 grille and Pontiac nameplate, this blacked out grille is for a model with the sport package.

1968

Visually the 1968 grille was the same as the 1967 model. But different part numbers were used. Once again the Firebird 400 models used a special grille. Front bumper nameplate and medallion all used the same part numbers, and restrictions.

1968 Grille Part Numbers

Model	Right-hand	Left-Hand
Except 400	9793725	9793726
Firebird 400	9793723	9793724

1969

An all-new look inhabited the 1969 Firebird. The grilles were restyled and featured a crosshatch pattern. Again two different grilles were used. The six cylinders and the 350-ci equipped models used grille assemblies part numbers 9795935 right-hand and 9795936 left-hand which were finished in argent silver. The Firebird 400, this includes the Trans Am models, used part numbers 546169 right-hand and 546170 left-hand and were painted flat black. It is incorrect for a six cylinder or 350 Firebird to use black painted grille assemblies or a Firebird 400 to use a silver painted grille assembly.

The Pontiac nameplate was stilled used on the left-hand side of the grille unit of all models including the Trans Am. The part number was the same as in two previous years. The medallion in the center of the front bumper was changed to part number 9750025.

Front bumper was also restyled and consisted of a two-part design. The inner portion , which supported the grille assemblies was chrome plated. All but those with a 400-ci V-8 used part number 9797454 while those with the 400-ci used part number 9795649. The outer portions of both bumpers were color keyed to the car. However, these bumpers were not just simply painted. Instead they featured a polyurethane material that covered the outer portion of the bumpers, which wrapped around the headlamps. This material was designed to withstand minor dings in the parking lot. Originally several different colors were available that matched the exterior of the car, as time went by a paint to match replacement panel part number 9784320 right-hand and 9784321 left-hand was offered. This panel appeared in September 1970 as the replacement part.

1969 grille, and Pontiac nameplate.

1969 Trans Am fender scoops

1969 Firebird front fenders featured simulated air scoops. They were used on all models except Trans Am.

1967-1968 Firebird nameplate, note the proper placement. Also the Firebird medallion was used on these models.

These nameplates were used on the front fenders of the 1969 Firebirds. Note the absences of the medallion that was used in earlier years.

Door and Door Hinges
1967

All doors are of one style, an open style. Only the 1967 door features a wing window design. Strap hinges were used, the upper hinges were a basic Pontiac part and listed as part number 7642529 and fit either side. While the bottom hinges were Tempest hinges and were unique to each side. The bottom hinges were listed as part numbers 7663424 right-hand and 7663425 left-hand. Hinges should be painted the same color as the exterior of the car.

Hoop style chrome plated door handles part numbers 5716870 right-hand and 5716871 left-hand were used. The button and shaft were listed as part number 9709600, and fit either side. Two different styles of interior door handles were used. Models with standard trim used part number 4468414, which can fit either side and are from the Tempest line. Those with Custom trim used paddle type handles listed as part numbers 7712035 right-hand and 7712036 left-hand. The door handle remote control unit should have a natural appearance two different sets were used. Those with standard trim used part numbers 4469683 right-hand and 4469684 left-hand, with custom trim part numbers 7660551 right-hand and 7660552 left-hand were used. The two are not interchangeable, and the proper interior door handle must be used.

The window regulator should have a dull aluminum finish. Two different styles were used. Those with standard trim used part number 7681466 right-hand and 7681467 left-hand, while those with custom trim used part numbers 7651903 right-hand and 7651904 left-hand. The window crank handle featured a plastic knob that was color-keyed to the interior trim. Note: there was a difference between those with standard trim and those with custom trim.

1967 Window Regulator Handles

Trim Color	Standard	Custom
Black	7680653	7680654
Red	7689668	7689653
Parchment	7689670	7689654
Gold	7689673	7689657
Blue	7689676	7719282*

* Bright Blue

The window wing handle was listed as part numbers 9708670 right-hand and 9708671 left-hand. A small red-orange color nylon bearing part number 4506696 was used between the handle and the glass to keep the glass from getting scratched.

1968

There was a change in design on 1968 doors. Early models used a door that was designed for use with a round-headed mirror while the later models used a rectangular shaped mirror. Note that both the driver's door and passenger's sides were changed.

1968 Door Part Numbers

Style	Left-hand	Right-hand
Early	7742292	7779693
Late	7742292	7742293

Upper hinges were restyled and now fit only one side of the car. They were listed as part numbers 8736244 right-hand and 8736245-left-hand. The bottom hinge was also restyled and given new part numbers of 8736248 right-hand and 7722191 left-hand. Both the upper and lower hinges were used on both the early and late style doors.

The outside door handles were the same units that were used on 1967 models, and used the same part numbers. The button and shaft were also the same. The interior door handles were restyled and once again weregrouped according to trim selection. Those models with the standard trim used part numbers 7743520 right-hand and 7743521 left-hand, which was used on the Tempest models. Those with the custom trim used part numbers 7759094 right-hand

7759095 left-hand. Remote control assemblies were also grouped according to the trim selection. Those with the standard trim used part number 8789692 right-hand and 8789693 left-hand, which is also in the Tempest line, while those with the custom trim used part numbers 8726277 right-hand and 8726278 left-hand. Remote controls from a car with standard trim will not fit a handle with custom trim or vice versa.
As in 1967 the window regulator unit was also used according to the trim selection. Those with standard trim used part numbers 7740411 right-hand and 7740412 left-hand, while those with custom trim used part numbers 7726801 right-hand and 7726802 left-hand. The difference in these units, is the location of the control knob, the two are not interchangeable, so be careful when swapping used doors to make sure you have the proper window regulator for your Firebird interior trim selection.

A chrome plated handle with a color-keyed knob that matches the interior trim was used on both the standard and custom trim models.

1968 Window Regulator Handle Part Numbers

Color	Part Number
Black	7752485
Red	7767319
Turquoise	7767320
Blue	7767322
Teal	7767323
Gold	7767325
Parchment	7767326

1969

Door was again restyled and the locking knob position changed, and will not fit earlier model years. The doors were listed as part number 8723785 right-hand and 8723786 left-hand. Upper hinge assemblies were new and given part numbers 8736246 right-hand and 8736247 left-hand. Lower hinges were also restyled and given part numbers 9817408 right-hand and 9817409 left-hand. Hinges should be painted the same color as the lower portion of the exterior of the car. The hoop style door handles were restyled and given part numbers 9706522 right-hand and 9706523 left-hand. The button and shaft was now listed as part number 9719686 and fit either door. The remote control assembly those with standard trim used part numbers 8789692 right-hand and 8789693 left-hand, while those with custom trim used the same part numbers as those in 1968 with custom trim. The interior door handle was the same exact parts and used the same restrictions as the 1968 models. The window regulators also used the same part numbers as in 1968. However, the window regulator handle was changed to part number 8732961 for all interior trim colors.

Door Latches and Strikers

A latch assembly was made from zinc-plated steel. Each model year used a different part. The door striker bolt was part number 4472299 all three years. It should have a flat black appearance. The striker latch plate was listed as part number 4469685 and should have a natural appearance.

Door Latch Part Numbers

Model Year	Part Numbers
1967	7648724-right hand 7648725-left-hand
1968	7740484- right-hand 7740485-left-hand
1969	8720401- right-hand 8720402 -left-hand

Outside Mirror-Standard

1967

A round head mirror part number 3909197 was standard on the driver's side door. A remote control mirror part number 3899857 was optional. A right-hand side outdoor mirror was optional as a dealer-installed option under part number 988596.It matched the other mirror in design, and was available on all Pontiac models.

1967 standard door mirror

1968

As stated previously early models used a round head mirror listed as part number 9787901, which is the same unit used on the 1967-68 Tempest models. Later built Firebirds used a mirror part number 3914753 with rectangular shaped head. It is important to match up the proper door and mirror. It is possible that the later built Firebird can use a round-headed mirror. This is due to the fact that a right-hand mirror was available. This mirror featured a round head shaped mirror and to make them match some dealers would pull the factory mirror and place a round-headed mirror on the driver's door. However, it is possible that a rectangular head mirror can be used on the driver's door and a round shaped mirror on the passenger as a dealer installed option.

1968 round head mirror was used only on early built 1968 Firebirds.

1969

All mirrors should have a rectangular shaped head. The mirror was listed as part number 9782390 and was used on all Pontiac models. Also available this year was a dealer installed mirror package. It too used the same mirror, but featured a matching mirror on the passenger' side door.

Late style 1968 side mirror.

1969 Standard outside mirror

Outside Mirror, Remote Control

1967
Listed as RPO D33 a round head mirror part number 3899857 was available on all 1967 Firebird models. A matching right hand mirror was not available from the factory but was available as a dealer installed option; see standard mirror section for more details.

1968
Just as with the standard mirror the optional remote control mirror also came in two forms. The early models used a round headed mirror listed as part number 9792565 while the later style used part number 3934583. Again a matching passenger side mirror was available only as a dealer installed option.

1969
The remote control mirror was available as either a factory option listed as RPO D33 , or as a dealer installed package part number 988642. Both packages used the same part number, which was listed as part number 9797263. When it was installed as part of a dealer installed option it usually included a passenger side mirror. Most dealer-installed mirror were placed on the base Firebird six cylinder and V-8 models.

Rocker Panels and Moldings

The rocker panel consisted of a two part design and inner panel and outer panel. The part numbers depend on the body style, as the convertible used different panels than the hardtop coupe. Rocker panel moldings were used on all 1967-1969 regardless of model, except the Trans Am. The Sprint models required special rocker molding to support the medallion. This medallion was listed as part numbers 9788227 right-hand and 9788228 left-hand, in 1967. In 1968 the medallion was changed to part number 9785957 and would fit either side of the car, and continued to be used on the 1969 models.

Rocker Panel Part Numbers

Body Style	Inner	Outer
Hardtop	9808988- R.H.	7741310- R.H.
	9808989- L.H.	7741311- L.H.
Convertible	7704362- R.H.	7727648- R.H.
	7704363- L.H.	7727649- L.H.

1967-1969 Rocker Panel Moldings

Model Year	Engine	R.H.	L.H.
1967	230-1bb or V-8	9785876	9785877
1967	230-ci 4-bbl	9785878	9785879
1968	250-1bb or V-8	9785876	9785877
1968	250-ci 4-bbl	9792936	9792937
1969	250-1bb or V-8	9794767	9794768
1969	250-ci 4-bbl	9794784	9794785

Roof and Drip Moldings 1967-1969

The roof panel was listed as part number 7585083 all three years. The windshield moldings were constant throughout the three-year span, but were dependent on the body style.

Windshield Molding Part Numbers

Body Style	Upper	Sides	Lower
Hardtop	7638868	7639558- R.H	7639926-R.H.
		7639559- L.H.	7639927-L.H.
Convertible	7638870	7639560- R.H.	76448736- R.H.
		7639561- L.H.	76448736 L.H.
Header Panel Moldings			
Convertible	Upper	7639786- R.H.	7639787- L.H.

Drip rails were used on the hardtop models only and are affected by whether or not the vinyl top was ordered. Those with a vinyl roof used different moldings than those without a vinyl roof.

Windshield Molding Part Numbers

Roof Drip	Scalp Molding (Roof Drip)	Pillar Drip
7637324- R.H.	7639406- R.H.	7656807-R.H.
7637325-L.H.	7639407- L.H.	7656808- R.H.

Scalp Molding Pillar Drip
7661072- R.H 7661073-L.H.

The rear glass moldings were used only on the hardtop model. The upper molding was listed as part number 8775869 all three years and was held with snap on clips or retainer, which was attached to the panel. The side moldings were listed as part numbers 8775872 right-hand and 8775873 left-hand. They were held in place with snap clips that snapped on to the moldings. The lower molding was a two piece design listed as part numbers 7639866 right-hand and 8775875- left-hand. This molding attached the same way as the upper molding did.

Rear Quarter Panel And Moldings 1967-1969

The rear quarter panel was dependent on the body style as the hardtop used a different panel than the convertible did. If a vinyl top or two-tone was ordered on the hardtop than a quarter belt molding part numbers 3914078- right-hand and 3914079- left-hand was used. Quarter pinch molding part numbers 3914092 right-hand and 3914093- left-hand were used on the convertible models. While the above part numbers remained unchanged during the three year run that this book covers, the wheel well opening trim is different. The 1967-68 models used part numbers 4229493 right-hand and 4229494-left-hand but the trim was restyled in 1969 to fit the new quarter panels. These moldings are listed as part numbers 8700967 right-hand and 8700969 left-hand. All three years the trim was held in place with Phillips head screws.

The 1967-1968 Firebirds featured simulated louvers in the rear quarter panel, in front of the rear wheel opening. Early 1967 models used chrome moldings that feature a black stripe. These moldings were listed as part number 7795005, mid way through the 1967 model year run these molding were replaced with all chrome trim listed as part number 4229997, which was used on all 1968 models. It is incorrect for a 1968 model to have the black accented chrome trim. However most reproductions are of the solid chrome type and this would be considered incorrect for an early built 1967 Firebird.

Rear Quarter Panels Part Numbers

Model Year	Body Style	R.H.	L.H.
1967	Hardtop	7748126	7748127
1967	Convertible	7702812	770283
1968	Hardtop	7738876	7738877
1968	Convertible	7741303	7741304
1969	Hardtop	8784180	8784181
1969	Convertible	7779482	7779483

Sprint models used special medallions on the rocker panel moldings. Shown here a medallion for a 1967 model.

Early 1967 Firebird used black accented trim on the rear quarter panel. Shown here is the late style.

Early style 1967 rear end panel gas cap and medallion.

Late 1967-1968 rear end panel gas cap and medallion.

Trunk Lid and Rear End Panel

The trunk lid used part number 8783521 all three years for all body styles. Likewise only one set of hinges were used and listed as part numbers 9709424 right-hand and 9709425 left-hand. Two different sets of torque rods were used. Part numbers 7661531 right-hand and 7661532 left-hand were used on all 1967 and 1968 models and again in 1969, except those with a rear spoiler. The heavy rear spoiler required special heavy-duty torque rods that were listed as part numbers 7733799 right-hand and 8780759 left-hand.

The rear end panel supported the tail lamps, and trunk lock mechanism. The 1967 and 1968 models used part number 7722988 both years. However be aware of a change in the fuel door in 1967. Early models used part number 7702794, which features a fingertip lift, while later 1967 and all 1968 Firebirds used a flat lid listed as part number 7770999. The Firebird medallion was also changed. Early models (with fingertip lift lid) used a smaller bird medallion listed as part number 9789710, while the later style (flat lid) used part number 9785962. The Pontiac name was spelled in individual letters high up on the rear end panel, just above the cylinder lock assembly. A natural appearing lock assembly part number 4753019 was used, the striker plate was listed as part number 7712075 and was attached to the underside of the trunk lid.

1967-1969 Rear End Panel Letters

P	7712137	T	7712140
O	7712138	I	7712411
N	7712139	A	7712142
		C	7712143

The rear end panel was restyled in 1969 and given part number 8716471. The fuel door was deleted, and the fill tube was moved down under the license plate. A Firebird medallion part number 8742310, Pontiac name was placed below this medallion. The part numbers were the same as they were in 1968.

A rear spoiler was used only the 1969 Trans Am as standard equipment, it was listed as part number 546530. In addition to the heavy-duty torque rods, a blue colored TRANS AM decal part number 546289 was placed on the top face of the rear spoiler in the center.

The rear end panel should be painted the same color as the exterior of the car with one exception the 1969 Trans Am. Trans Am model should have the rear end panel painted the same color as the stripes. It is incorrect for a 1969 Trans Am to have a have a white painted rear end panel, for more details see Sport Stripes this chapter.

1969 Firebird medallion was used around the trunk cylinder lock.

1967-1968 rear end panel lettering location.

Even though the letters used the same part numbers the 1969 rear end panel lettering location was changed.

A combination of medallion and side lamp bezel was used on the rear quarters of the 1969 Firebirds.

1969 models with a 400-ci V-8 used callouts on the rear end panel.

Location of 400 rear end nameplate.

Glass

Model	Clear	Tinted	Notes:
Windshield			
Hardtop	7631031	7631032	1967-69
Convertible	7631033	7631034	1967-69
Vent Windows			
Hardtop/Convertible	9708666-R.H.	9708686- R.H.	1967 only
	9708667-L.H.	9708687- L.H.	
Door Glass			
Hardtop/Convertible	7636736- R.H.	7636738-R.H.	1967 only
	7636737- L.H.	7636739-L.H.	
Hardtop/Convertible	7777404- R.H.	7777400-R.H.	1968-1969
	7777405 L.H.	7777401- L.H.	
Quarter Glass			
Hardtop	7707436- R.H.	7707442- R.H.	1967 only
	7707437- L.H.	7707443- L.H.	
Convertible	7707448- R.H.	7707454- R.H.	1967 only
	7707449- L.H.	7707455- L.H.	
Hardtop	7720356- R.H.	7720358-R.H.	1968-69
	7720357- L.H.	7720359- L.H.	
Convertible	7720360- R.H.	7720362- R.H.	1968-69
	7720361- L.H.	7720363- L.H.	
Back Glass			
Hardtop	7600989	7600990	1967-69

Quarter Window

Quarter glass is unique for the 1967 models and will not fit other model years. Also be aware that there is difference between those on a hardtop and the glass used on the convertible. The window regulators should have a natural appearance.

Quarter Glass Window Regulators

Model Year	Right-hand	Left-hand
1967	7660238	7660239
1968-69	7777450	7777451

The window regulator handle was bright plated and featured a color-keyed to the interior knob for 1967 and 1968, for 1969 part number 8732962 was used with all interior trim color. Note that this was a special handle and those from Tempest or full-size models will not fit. Those from a 1969 Grand Prix will interchange.

1967 Window Regulator Handles Quarter Window

Trim Color	Standard	Custom
Black	7680653	7680654
Red	7689668	7689653
Parchment	7689670	7689654
Gold	7689673	7689657
Blue	7689676	7719273*

* Bright Blue

1968 Window Regulator Handle Quarter Window

Color	Part Number
Black	7752485
Red	7767319
Turquoise	7767320
Blue	7767322
Teal	7767323
Gold	7767325
Parchment	7767326

Glass Date Coding

It is important that the proper date code,be on the glass for your Firebird to be considered 100 percent correct. Most glass for early generation Firebirds was manufactured by the Libby Owens Ford Manufacturing, didn't know you had Ford products in your Pontiac did ya? The other manufacture was the Pittsburg Plate Glass. The plant code will be etched into the glass, LOF for Libby Owens Ford, and PPG for Pittsburg Plate Glass.

General Motor's trade name for the glass will appear on the top line. These include: Soft-Ray, Shaded, Plate and Flo-lite. Soft-Ray is the most common and is the name used for tinted-glass. The term Shaded will most likely appear on a windshield, and means that only the top portion of the glass is tinted. Plate or Safety Plate is the term for standard non-tinted glass. Flo-lite is rare but describes the method that the glass was manufactured.

The code AS1 or AS2 are the American Standard codes that are set forth by the Society of Automotive Engineers. The code AS1 stands for laminated safety glass, which is used in the construction of windshields. The code AS2 is for Solid Tempered Glass. You will see these codes along with the name that appear on the glass. The DOT code will appear on some late built and 1969 models. Also appearing near the bottom line of information is the manufactures code for the type of glass.

The date will be the most important bit of information that can be on your glass. For a car to be considered a numbers matching car, the proper date code must appear on ALL the glass. The date code should not be later than the final assembly date of your car. In fact it is common for the glass to be built a month to two months before your car.

The date code will appear following the American Standards code and the Assembly plant code. It will consist of a two letter code that will indicate the month and the manufacture year. Note that the year code and the model year may not always agree with each other. A code for August 1967 would not be for a 1967 model but instead a 1968 model. While a code for May 1968 would also be for a 1968 model. There may be a code that follows the date code, but do not confuse this with the date code; it is a code that represents the assembly plant location.

There were two type of glass options available for your Firebird. RPO A01 was for all tinted glass while RPO A02 was for just a tinted windshield. Tinted glass was not a dealer-installed option. And though most cars with air conditioning came with all tinted glass it was not a required option.

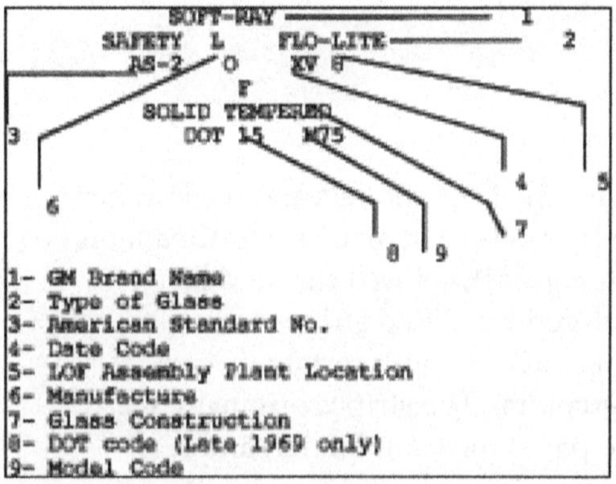

Breakdown of a typical glass code

Manufactures Model Numbers

Model Code	Type of Glass
M3	Non-tinted windshield
M4	Tinted windshield
M51	Non-tinted door glass
M52	Non-tinted quarter
M55	Tinted back and vent
M71	Non-tinted door glass
M72	Non-tinted quarter and back
M74	Tinted door glass
M75	Tinted quarter ,vent and back glass
M78	Tinted Quarter glass

Month	Code	Month	Code	Month	Code	Month	Code
Jan.	N	Feb.	X	March	L	April	G
May	J	June	I	July	U	Aug.	T
Sept.	A	Oct.	Y	Nov.	C	Dec.	V

Year of Manufacture

1966*	1967	1968	1969
A	Z	X	V

*-July to Dec. only

LOF Plant Locations

Code	Plant	Type of Glass
6	Toledo, OH	Tempered glass
8	Toledo, OH	All
10	Lathrope, CA	Tempered glass

Glass date codes

Rear Bumpers

Only two different rear bumpers were used throughout the three-year run that this guide covers. Part number 9788872 was used on 1967 and 1968 models while part number 9795392 was used on the 1969 models. The two bumpers are not interchangeable.

The 1967-68 bumper used a intermediate bar part numbers 9789295 right-hand and 9789296 left-hand, to attach it it to the body of the car along with outer bars listed as part numbers 9789861 right-hand and 9789862 left-hand, and a center bracket part number 9789706. All of these components should be painted flat black.

The 1969 bumper used brackets to attach the bumper to the car. All brackets should be painted flat black.

1969 Rear Bumper Brackets

Outer	9796004- R.H.	9796005- L.H.
Intermediate	9795797- R.H.	9795798- L.H.
Center	9796123	

Sport Stripes
1967-1968

Basically there was one stripe treatment with the three different styles. The Rally stripe was routed along the middle of the car from the rear edge of the front wheel well opening lip to the front edge of the rear wheel well opening lip. The stripe pattern was of a wider stripe flanked on both sides by narrower stripes. There was a set of stripes for the H.O. models and the regular Firebird models. The stripes were available in white for dark colored cars and black for light colored cars. Red stripes were also available but only with certain trim or exterior colors. The stripes originally were in three parts; the front fender, and rear quarter panel.

1967-1968 Rally Stripes

Color	Fender	Door	Quarter Panel
Black	9789312- R.H.		
	9789313- L.H.		
Red	9789316- R.H.		
	9789317- L.H.		
White	9789314- R.H.		
	9789315- L.H.		

1967-1968 Rally Stripes For H.O. Cars

Color	Fender	Door	Quarter Panel
Black	9789825- R.H.		
	9789826- L.H.		
Red	9789829- R.H.		
	9789830- L.H.		
White	9789827- R.H.		
	9789828- L.H.		

H.O. Logo on front fender stripe

1968 Exterior Color Stripe color Combinations

Color	Stripe Color	Trim Combinations
Starlight Black	White	Black, Parchment, Gold, Turquoise, Teal
		Blue
	Red	Red
Cameo White	Black	Black, Parchment, Teal, Turquoise
		Gold
		Blue
	Red	Red
Alpine Blue	White	Parchment
		Blue
	Black	Black
Agenda Blue	White	Blue, Parchment
		Black
Nordic Blue	Black	Black
	White	Blue
		Parchment
April Gold	Black	Black
	White	Parchment
		Gold
Meridian Turquoise	Black	Black
	White	Parchment
		Turquoise
Aleutian Blue	White	Blue
		Black
		Parchment
Flambeau Burgundy	White	Black
		Parchment
Silvermist Green	Black	Black
	White	Parchemnt
Solar Red	Black	Black
	White	Parchment
Primavera Beige	Black	Gold
		Black
		Parchment
Nightshade Green	White	Black
		Parchment
Mayfair Maize	Black	Black
		Parchment

1969

There were no optional tape stripes for the Firebird this year. All stripes were painted on. The Trans Am used two racing stripes that ran from the edge of the twin hood scoops up and over the roof and down the deck lid. Some models have the stripes over the rear spoiler others did not. Only blue stripes were used.

Cordova Top

1967-1969

A vinyl top was optional on all hardtop Firebirds all three years. However, color choice varies with the model year. For 1967 the choices were black, white, or cream. For 1968 black and white return but gold and teal were added and cream was deleted. For 1969 the color choices were black, white, parchment, dark blue, dark green and fawn. Color choice depended on the exterior color. Black or white Cordova tops were available with any exterior color. The others had restrictions for recommended color choices.

The proper color code for the Cordova top can be determined by look at the upper body color code on the Fisher Body tag (see Chapter 1 for detail on the location of this tag). The following codes were used.

Cordova Vinyl Top Color Codes

1967	
Code	**Color**
1	White
2	Black
7	Cream
1968	
1	White
2	Black
5	Teal
8	Gold
1969	
2	Black
3	Dark Blue
5	Parchment
8	Dark Fawn
9	Dark Green

Folding Top

1967-1969

The convertible top consisted of a frame network of rails. The front rail was listed as part number 7702456. The side link rail was listed as part numbers 7639080 right-hand and 7639081 left-hand for all three years. The rails should be painted 60-degrees gloss black. The rear rails were listed as part numbers 7639074 right-hand and 7639075 left-hand all three years; they too should be painted 60 degrees gloss black. The center rail depended on the type of top ordered. A manual top was standard and it used part numbers 7639060 right-hand and 7639061 left-hand. A power top was optional and used part numbers 7639062 right-hand and 7639033 left-hand. The center rail should also be painted gloss black. The lock part 7664618 right-hand and 7664609 left-hand were painted.

The convertible top came in a variety of colors and the color choices are dependent on the model year. For 1967 black, white, blue, turquoise or cream were available. For 1968 the color choices were black, white, gold or teal. In 1969 black, white blue or green were offered. All exterior colors could be ordered with a black or white convertible top.

A little know and very rare option was the dealer installed tonneau cover that was available as a dealer installed option for 1967 Firebird convertibles. Basically it was a slip of canvas that cover a hole that was cut into the convertible top, sort of like later Ventura II model sun roofs.

The power top switch assembly was listed as part number 3840095 in 1967 and 1968. Gray and purple colored wires were routed from the switch to the motor/pump assembly listed as part number 7710923. In 1969 the switch was changed to part number 9797081 which is the same switch used in all Pontiac convertibles. The motor/pump assembly was the same unit that was the previous two years.

A 108-in. long hose part number 4881617 was routed from the pump to the lift cylinders. Two different sets of lift cylinders were used. Those with manual top were counter balanced and used part number 4412503 while those with the power top used part number 7704518; the lower hose was listed as part number 7794131. These part numbers were used all three years.

Weather Strip

Front Roof Rail

1967				
Black	**White**	**Blue**	**Turquoise**	**Cream**
7665021	7665022	7665023	7715359	7715360
1968				
Black	**White**	**Gold**	**Teal**	**Rubber**
7665021	7665022	7773053	7715359	772255
1969				
Black	**White**	**Blue**	**Green**	**Rubber**
7665021	7665022	874799	874800	772255

Folding Rail, Front

7722080- R.H. 7722081 LH

Roof Rail, Center

7645432-R.H. 7645433-L.H

Roof Rail, Rear

7645434-R.H. 7645435-L.H.

Fuel Tank and Cap
1967-1968

An 18-1/2 gallon tank was used on all 1967 and 1968 Firebird models. The tank should have a natural unpainted steel appearance. Two support straps part number 3908341 was used to support the tank. The in tank fuel gauge assembly varied according to the engine usage. All models used an anti-surge vented gas cap listed as part number 9794205 both years.

In Tank Fuel Gauge Part Numbers

230-ci 1-bbl	6428069
230-ci 4-bbl	6428069*
	6427584**
250-ci 1/bbl/4-bbl	6428069
326-ci 2-bbl/4-bbl	6427255*
	6427857**
350-ci 2-bbl	6427255*
	6427857**
350-ci 4-bbl	6427857
400-ci	6427857

1969

Two different tanks were used. Those with a vented system and those with a non-vented system. The straps and in tank gauges remained the same. Caps were listed as part number 3931449 with a vented system and part number 9790483 with a non-vented system.

Chapter 9

Interior

Instrument Panel
1967

The instrument panel was listed as part number 7653486 for all models. The instrument panel was painted in a color that was color-key or contrasted to the interior trim color. The upper portion of the instrument panel was painted a 0 degrees gloss finish while the lower portion was painted in a 60-degrees gloss finish. The glove box compartment was made of black cardboard and was held together with ten staples. It was held to the instrument panel with natural appearing Phillips head screws and u-nuts. Part number 3891671 was used without air conditioning, and part number 3891766 was used with air conditioning. The air conditioning unit has a more sloped back top to provide clearance for the air duct hoses. For this reason a standard compartment will not fit a car with air conditioning. The door assembly part number 9777956, should be painted to match the interior trim in a 60-degree gloss except with the parchment trim, where it should be painted 60-degrees gloss black.

A crash pad matched the dash in color. If no radio was ordered than a blank panel part number 3893845 was used over the opening. The ashtray 3891667 was used in all models; the outer portion of the tray should be painted to match the instrument panel.

1967 Instrument Panel Color Part Numbers

Codes	Color	Upper Panel (Flat)			Lower 60-degree Goss		
		Dupont	Ditzler	Rinshed-Mason	Dupont	Ditzler	Rinshed-Mason
272,269,260 254,272,269	Black	4428-L	9317	400-A-946	88	9248	400-A-946
251,2572,71 267	Gold	9376-L	22837	67B72	9390-L	22839	67B78
252,258	Red	9376-L	22837	67B72	9366-LM	71591	67B57R
256	Turquoise	9374-LH	13366	67B31	9399-LH	43680	67838
255, 265	Bright Blue	9286-LH	66B27	DIA13216	9271-LH	67024	13223
250,270	Blue	9373-LH	13365	67828	9395-LH	13383	67B29
281	Leather Brown	9583L	23143	169C77	9549L	23160	169C75

1968 Instrument Panel Color Part Numbers

Codes	Color	Upper Panel (Flat)			Lower 30-degree Goss		
		Dupont	Ditzler	Rinshed-Mason	Dupont	Ditzler	Rinshed-Mason
272,253,268 259, 273,262, 275,260	Black	4428-L	9317	400-A-946	99-L	9266	170B40
250,255	Blue	9453-LH	13556	168B22	9467-L	13567	168B29
251,257	Gold		43799		9478-L	22988	168B74
258, 252	Red	9201_lm	71486	65B55M	9366-LH	71591	167B851R
261,256	Turquoise	9456-L	43799	168B71	9771-LH	13570	168B35
250,270 281	Blue Saddle	9373-LH	13365	67828	9395-LH	13383	67B29

1968

Though it looked similar to the 1967 version the 1968 instrument panel used a different part number. Listed as part number 7792989, it used the same glove box compartments and doors as before with the same restrictions.

The upper portion of the panel was still done in a flat finish, while the lower are now done in 30-degree gloss. It is important that the proper finish is applied for correctness. Once again color matched the interior trim, except with the white (parchment) interior trim, where the instrument panel was painted black.

1969

The instrument panel was all new for 1969 and used part number 8716459. Though it still used the same glove box compartments as the 1967-68 models did, the glove box door was changed. Listed as part number 9797557 it will not fit earlier models. It has no provision for a Firebird medallion. The radio cover panel was changed to part number 9796880. Painting instructions are the same as in 1968.

Ashtray was listed as part number 6288831 and used a fiber optic guide, this receiver actually appeared in late built 1968 Firebirds.

1969 Instrument Panel Color Part Numbers

Codes	Color	Upper Panel (Flat)			Lower 30-degree Goss		
		Dupont	Ditzler	Rinshed-Mason	Dupont	Ditzler	Rinshed-Mason
227, 207, 217 228, 208, 218	Black	4428-L	9317	400-A-946	99-L	9266	170B40
202, 212, 293	Gold	9583L	23143	169C77	9549L	23160	169C75
206, 216	Green	9582LH	43993	16921636	9593LH	4410	169V35
214	Red	9580LM	71643	169B63M	9528LH	71751	168C51R
261,256	Turquoise	9456-L	43799	168B71	9771-LH	13570	168B35
200,210	Blue	9578LH	13767	169V22	9586LH	13789	169B29

Instrumentation and Controls
1967-1968

Laid out directly in front of the driver were two round pods. To the far left was a 120-mph speedometer. Certain options like the rally cluster, which included auxiliary gauges, and tachometer, and the Safeguard, which was a buzzer that would go off when the vehicle hit a preset speed that was set by the owner, affected the speedometer usage. Just because your car has a hood mounted tachometer do not assume that your car came with the Rally cluster. The hood-mounted tachometer was a separate option and available with or without the rally cluster. The standard speedometer pod also housed the parking brake warning lamp, left-hand turn signal, and oil pressure warning lamp.

The 1967 Firebird used the same panel and instrumentation that was used in the Chevrolet Camaro.

1967-1968 Speedometer Part Numbers

Option	Part Number
1967	
Standard	6480794
Safeguard	6481283
With Rally Cluster W/O Safeguard	6480796
With Rally Cluster With Safeguard	641254
1968	
Standard	6481931
Safeguard	6481932
With Rally Cluster W/O Safeguard	6481933
With Rally Cluster With Safeguard	641934

1967-1968 Instrument Cluster Printed Circuit

Option	Part Number
Standard	6290070
With Rally Cluster	6293188

1967-1968 Fuel Gauge

Option	Part Number
1967	
Standard with backing plate	6480802
With Rally Cluster	6430659
1968	
Standard with backing plate	6431007
With Rally Cluster	6430930

To the right of the speedometer was the fuel gauge. Also housed in this pod was the right-hand turn signal indicator and generator light. A printed circuit board controlled the two pods.

When the optional 1968 rally cluster was ordered a 160-mph speedometer was used. The fuel gauge was displaced with a different unit that contained gauge units for the generator, water temperature, oil pressure and a smaller fuel gauge.

1967-1968 Oil Pressure Gauge

Option	Part Number
1967	6461289
1968	6461552

To the far left-hand side were to the controls for the wiper and lamp switch. To the far right-hand side was the lighter assembly and ignition switch. For more information on these controls see the electrical chapter of this guide.

1969

Revamped system with a three-pod design. Main gauges were laid out in front of the driver. To the left was the fuel and temperature gauge, if ordered with auxiliary gauges. To the right and in the center was the 120-mph speedometer. To the far right was a blank panel with the Pontiac logo.

As before options like Safeguard and the Rally cluster affected the instrument cluster usage. Two rally clusters were available RPO W63 was with a clock and RPO U30 was with an in dash tachometer. The clock or tachometer displaced the blank panel to the far right-hand side. RPO W63 was available with the hood tachometer, but RPO U30 was not. For more information on the hood mounted tachometer see the options chapter of this guide.

1969 Instruments

Standard	Speed Guard	Rally	Notes
Speedometer			
6492260	6492261	6492260* 6492261**	*-w/o Speed guard **-With Speed guard
Fuel Gauge			
6431093	6431093	6431155 6431083#	#-Auxiliary gauges
Oil Pressure			
N/A	N/A	6461987	
Water Temperature			
N/A	N/A	6489331 6489469#	

Steering Wheel
1967

Standard steering wheel was a three spoke unit that was molded in colored plastic that matched the interior, except for white, which was molded in black plastic. In the center of the steering wheel was horn cap listed as part number 9787528, and featured the Pontiac logo.

The deluxe steering wheel looked similar to the standard unit in design, but has provisions for deluxe trim on the spokes that included bright horn buttons. The wheel was standard in models with the custom trim. The deluxe wheel was also a separate option.

Early models used a 16-in diameter steering wheel, while later models used a 16-1/2 in. diameter steering wheel. There is no date of this change but it is believed to be around the end of the 1966 or the beginning of 1967. It would be incorrect for an early built Firebird to have a 16-1/2 in. diameter wheel. Note that this applies to both the standard and deluxe steering wheel.

The Custom sport steering wheel was a popular option. It featured a simulated wood rim and was listed as part number 9787429. The custom sport center cap should have a polished chrome appearance. In 1970 Pontiac deleted this cap and integrated it with the 1968 part, which has a brushed appearance. While they will fit, a brushed center cap is incorrect for the 1967 models.

1967 Deluxe steering wheel.

1967 Standard and Deluxe Steering Wheels

Trim Code	Wheel Color	Steering Wheel Part Numbers Standard	Deluxe	Notes
272,269,260	Black	9746294*	9746314*	*-16-in
254,272,269		9745779	9746302	
251,2572,71	Gold	9746297*	9746317*	*-16-in
267		9748549	9746307	
252,258	Red	9746298*	9746318*	*-16-in
		9745784	9746308	
256	Turquoise	9746396*	9746316*	*-16-in
		9745782	9746306	
200,210	Blue	9746295*	9746315*	*-16-in
		9745781	9746304	
255, 265	Tryol Blue	N/A	9748452*	*-16-in
			9746310	

1968

Design wise the standard steering wheel was the same unit as in late 1967. All steering wheels measured 16-1/2 in. diameter. Again the wheels were color keyed to the interior trim, except when white trim was used then a black colored steering wheel was used. The center cap was changed to part number 9793037 and contains the Pontiac logo.

Deluxe steering wheel continued as an option and was standard with custom trim. The steering wheels were color-keyed to the interior trim, except for parchment, which was molded in black. The design was similar to that used in late 1967 but the horn buttons were color-keyed to match the steering wheel.

The Custom sport steering wheel continued, but was redesigned. It was much shallower than the year before. Two different wheels were used. All factory models used part number 5464470. It was reported in Dealer Campaign Bulletin number 68-C-4 reported problems with this wheel -one or more of the spokes can break off at the hub. Thus the wheel was recalled and part number 546470 took its place. It is reported that some of these wheels were not replaced. Either wheel would be considered correct. For safety part number 5464470 should not be used on a driver. The center cap part number 9787953 looked similar to the 1967 but had a brushed texture instead of the bright polished texture that was used in 1967. It would be considered incorrect for a 1968 Firebird to have a polished center horn cap.

Horn button Part Numbers

Black	9748522
Red	9748523
Blue	9748524
Gold	9748525
Turquoise	9748526
Saddle	9750329

1968 Standard and Deluxe Steering Wheels

Trim Code	Wheel Color	Steering Wheel Part Numbers Standard	Deluxe	Notes
272,269,260 254,272,269	Black	9745779	9748532	
251,257, 271 267	Gold	9748541	9748535	
252,258	Red	9745784	9748533	
256	Turquoise	9748542	9748536	
200,210	Blue	9748540	9748534	
281	Saddle	N/A	9750124	

1967 custom sport steering wheel featured a simulated wood rim

1969

All new steering wheels graced the 1969 Firebird line. All models were standard with the two spoke standard steering wheel. As in years past the steering wheel was molded in colored plastic that matched the interior trim, except for white, which used a black colored steering wheel. The horn shroud matched the steering wheel in color and featured the Pontiac symbol in the middle.

1969 Standard Steering Wheel Center Cap

Black	9796889
Gold	9797360
Green	9797387
Red	9797364
Turquoise	9797365
Blue	9797158

The Deluxe steering wheel was available as a separate option, RPO N30, or as part of the custom interior group. It used a three-spoke design, which was color keyed to the interior trim, except white, which used a black colored wheel. The center cover matched the wheel in color and featured soft touch vinyl horn buttons that too matched the steering wheel in color.

1969 Custom Sport steering wheel.

1969 Deluxe steering wheel, is the most common found on 1969 models

1969 Deluxe Steering Wheel Center Cap

Black	9797158
Gold	9797384
Green	9797387
Red	9797388
Turquoise	
Blue	9797389

1969 Standard and Deluxe Steering Wheels

Trim Code	Wheel Color	Steering Wheel Part Numbers		Notes
		Standard	Deluxe	
227, 207, 217	Black	9751053	9751059	
228, 208, 218				
202, 212, 293	Gold	9751055	9751061	
206, 216	Green	9751054	9751060	
214	Red	9750334	9750341	
261,256	Turquoise	9748550	N/A	
200,210	Blue	9750336	9751064	

1969 Custom Sport Wheel Center Cap

Black	9797101
Gold	9797481
Green	9797484
Red	9797485
Turquoise	9797482
Blue	9797486

The custom sport wheel (part number 9795885 RPO N34) was also restyled. The rim was simulated wood and featured a colored horn bezel that matched the interior trim, except for white, which used a black center cap.

Available on the 1969 Trans Am was a special Formula type steering wheel. Listed as part number 546296, it featured leather wrapped rim with a black center cap and brushed center spokes with two holes in each spoke. This wheel was most common on the 1970 Trans Am, but was available in all Pontiac models under option code RPO NK3.

Seats

1967

All vinyl bucket seats were standard in all Firebird models, including the base six cylinder. The insert consisted of a pattern of multiple horizontal pleats done in Madrid grain. They were available in either standard or custom trim, the seat pattern did not change if the custom trim was ordered. Color choices were black, bright blue, turquoise, dark blue, gold, parchment or red, with bright blue being available only with custom trim.

Available as an option for the hardtop coupe only was a full width bench seat. Done in the same grain and pattern as the bucket seats this option was not that popular, and only a handful of Firebirds were made with a bench seat. Color choices were limited to black, bright blue, dark blue, gold and parchment. Want a rarity? Try a Firebird with a bench seat and custom trim; yes, this set up was available, but less than a dozen were believed to be built this way.

The seat frame and springs should be painted 60-degrees gloss black. The seat back to the cushion hinges should have a natural unpainted metal appearance, which can be duplicated by painting with a finish that will duplicate this look. Several different manufactures offer this including Eastwoods. The hinges were covered with injection-molded plastic covers that had a pebble-grain finish and were molded in a color that matched the seat trim. Reproductions are usually in black, and should be painted to match the interior trim in a 60-degree gloss finish. The original parts were listed as:

1967 Seat Hinge Covers Driver's Bucket

Color	Inner	Outer
Black	7684911	7684910
Red	7684923	7684922
Dark Blue	7715421	*7715420*
Turquoise	7715835	*7715834*
Gold	7684916	*7684915*
Parchment	7761207	*7761206*

1967 Seat Hinge Covers Passengers Bucket

Color	Inner	Outer
Black	7684910	7684911
Red	7684922	7684923
Dark Blue	*7715420*	*7715421*
Turquoise	*7715834*	*7715835*
Gold	*7684915*	*7684916*
Parchment	*7761206*	*7761207*

1967 Seat Hinge Covers Bench Seats

Color	Inner	Outer
Black	7684911	7684910
Dark Blue	7715421	*7715420*
Gold	7684916	*7684915*
Parchment	7761207	*7761206*

The bucket seats used a protective ABS plastic cover. Originally these covers were molded in a color that matched the seat trim in color. Most reproduction parts are in black only, or used parts will most likely not match your (seat backs from 1968-70 fit) trim, but they can be painted to match your trim. Before applying paint the panel should be thoroughly cleaned using the proper solvent. Then lightly scuff the surface with #400 grit sandpaper. Then clean again with the solvent cleaner, and dry thoroughly, with a lint free cloth. Then coat with acrylic lacquer that matches the trim. Color part numbers can be found under the instrument panel heading use 60-degree gloss. Allow to drying completely then rub out and polish.

1967-1969 Bucket Seat Back Covers

1967

Color	Driver	Passengers
Black	7677996	7677995
Dark Blue	7715425	7715424
Turquoise	7715399	7715398
Red	7678009	7678008
Parchment	7678007	7678006
Gold	7678001	7678000
Bright Blue	7677998	7677997

1969

Color	Driver	Passengers
Black	7747487	7747486
Green	8730897	8730896
Blue	8730893	8730892
Gold	8730891	8730890
Parchment	8730899	8730898

1968

Color	Driver	Passengers
Black	7747487	7747486
Red	7747489	7747488
Gold	7747493	7747492
Parchment	7752444	7752443
Teal	7783413	7783412
Turquoise	7761799	7761798
Saddle	8712515	8712515

Saddle in 1968 is a late model release.

1968

All vinyl bucket seats were again standard in all Firebird models. However, this year the standard and custom trim used different insert patterns.

The standard trim looked like the 1967 seat trim with multiple horizontal pleats. It was available in black, bright blue, dark blue, turquoise, gold, parchment, and red. The custom trim used wide vertical pleats done in comfort weave vinyl with Madrid vinyl skirts, it was offered in the same colors as the standard trim. Bench seats remained as an option for the Firebird hardtop, but were available only in black or parchment colors.

You have seen it mentioned throughout this guide but what is the truth about interior trim code 281? It is an interior code that most Firebird books do not list as being available; it isn't listed being available on sales brochures. Interior trim code 281 was introduced late in the year; most research indicates it was released sometime during May 1968, it is a saddle color and was available, (as far as research shows) as a custom trim only and used leather inserts instead of vinyl. This interior trim also used a saddle color deluxe steering wheel; door panels and other interior trim components. If you have a Firebird with the trim code 281 and it was built before May 1968 please do a Fisher tag rubbing and send it to me in care of the publisher.

The seat frame and springs should be painted 60-degrees gloss black. The seat back to the cushion hinges should have a natural unpainted metal appearance. The hinges were covered with injection-molded plastic covers that had a pebble-grain finish and were molded in a color that matched the seat trim. Reproductions are usually in black, and should be painted to match the interior trim in a 30-degree gloss finish.

1967-1968 Standard seat trim.

1968 Seat Hinge Covers Driver's Bucket

Color	Inner	Outer
Black	7684911	7684910
Red	7684923	7684922
Turquoise	7761204	7761205
Teal	7783491	7783490
Gold	7731535	7731534
Parchment	7761207	7761206
Saddle	8712525	8712524

1968 Seat Hinge Covers Driver's Bucket

Color	Inner	Outer
Black	7684910	7684911
Red	7684922	7684923
Turquoise	7761205	7761204
Teal	7783490	7783491
Gold	7731534	7731535
Parchment	7761206	7761207
Saddle	8712524	8712525

1968 Seat Hinge Covers Bench Seat

Color	Inner	Outer
Black	7684911	7684910
Parchment	7761207	7761206

1969

All vinyl bucket seats were again standard in all Firebird models. The front bench seats remained an option for the hardtop coupe and was available in only black or parchment in standard trim only. Patterns and grain textures were the same as the 1968 models. Leather trim did return in custom trim bucket seat only, and in gold. Color choices were; black, parchment, blue, green, gold and red. All painting instructions are the same as in 1968.

1969 Seat Hinge Covers Driver's Bucket

Color	Inner	Outer
Black	7684911	7684910
Red	8721187	8721186
Blue	8721189	8721188
Gold	8756089	8756088
Parchment	8721193	8721192
Green	8726091	8712524

1969 Seat Hinge Covers Driver's Bucket

Color	Inner	Outer
Black	7684910	7684911
Red	8721186	8721187
Blue	8721188	8721189
Gold	8756088	8756089
Parchment	8721192	8721193
Green	8712524	8726091
Black	7684910	7684911

1969 Seat Hinge Covers Bench Seat

Color	Inner	Outer
Black	7684911	7684910
Parchment	8721193	8721192

196-1969 Deluxe seat trim.

All vinyl bucket seats were again standard in all Firebird models. The front bench seats remained an option for the hardtop coupe and were available in only black or parchment in standard trim only. Patterns and grain textures were the same as the 1968 models. Leather trim did return in custom trim bucket seat only and in gold. Color choices were; black, parchment, blue, green, gold and red. All painting instructions are the same as in 1968.

The kick panels were made from color molded plastic that matched the interior trim , except for parchment which used black. Two different seats of kick panels were used with each color, regardless of whether air conditioning was used or not .

Door and Kick Panels

1967

There is a physical difference between the standard trim and custom trim door panels, and they cannot be interchanged. The panels were covered in colored vinyl that matched the seat trim in color. The standard panels featured a pattern of multiple vertical pleats that were dielectrically pressed into Madrid grain vinyl. A separate armrest is used in the center of the panel. The panel is highlighted with two strips of chrome trim that cut across horizontally along the same axis as the armrest.

With the custom interior the door panels were more refined and had a Corvette feel to them.They featured a small patch of vertical pleats that were dielectrically sealed into the Madrid grain vinyl. The arm rests were built in and carpet, which matched the trim except for parchment which used black carpet, was placed along the bottom edge of the door panel.

1967 Kick Panels

Color	Drivers	Passengers
Without Air Conditioning		
Black	7664672	7664671
Red	7664674	7664673
Gold	7664676	7664675
Blue	7664678	7664677
With Air Conditioning		
Black	7664685	7664684
Red	7664695	7664694
Gold	7664687	7664686
Blue	7664689	7664688

1967 Arm Rest

Color	Drivers	Passengers
Standard Trim		
Black	4475429	4475429
Red	7654349	7654349
Gold	7654348	7654348
Blue	7713927	7713927
White	7732509	7732509
Bright Blue	7732510	7732510
Turquoise	7732511	7732511

1967 Door Panel Part Numbers

Interior Code and Color		Standard Door Trim		Custom Door Trim	
		Passenger	Driver	Passenger	Driver
250, 270	Blue	7714012	7714013		
251, 271	Med. Gold	7714014	7714015		
252	Red	7714016	7714017		
253. 272	Black	7714018	7714019		
254	White	7732514	7732515		
255, 265	Bright Blue			7721530	7721531
256	Turquoise			7714022	7714023
257, 267	Gold			7673666	7673667
258	Red			7673672	7673673
259. 269	Black			7673674	7673675
260	White			7714024	7714025

1968 Standard door trim panels. Note the Firebird nameplate in the center.

1968

While the seat trim remained visually the same from 1967, the door panels were drastically changed and will not fit earlier models, as there is no provision for the wing window control, as 1968 used a single one-piece side glass. The standard trim door panel featured twin chrome molding strips that went across the panel horizontally In between these strips were horizontal ribs that were sealed into the Madrid grain vinyl. Carpet that matched the door panel in color, (except for parchment which used black carpet), was placed below the bottom chrome strip molding. The Firebird nameplate part number 7728725, was used on this door panel.

Door Moldings with Standard Trim

	W/O Remote Control Mirror	With Remote Control Mirror
Upper	7737762	7726724-front 7759450-rear
Lower	7730732	7730732

The custom door panel had a sport car feel to it with a built in arm rest and pull handles that matched the door panel in color, these pull handles were the sane ones used on the instrument panel above the glove box which were part of the custom interior trim group. A bright molding part number 7743242, was placed just above the carpet that was routed on the bottom edge of the door panel. The carpet matched the door panel in color except for parchment, which used a black colored carpet. The Firebird medallion part number 7771090, was used on this door panel.

1968 Door Pull Handles

Color	Drivers	Passengers
Custom Trim		
Black	7770614	7770614
Red	7770616	7770616
Gold	7774100	7774100
Blue	7770620	7770620
White	7774101	7774101
Turquoise	7770617	7770617
Saddle	8712165	8712165

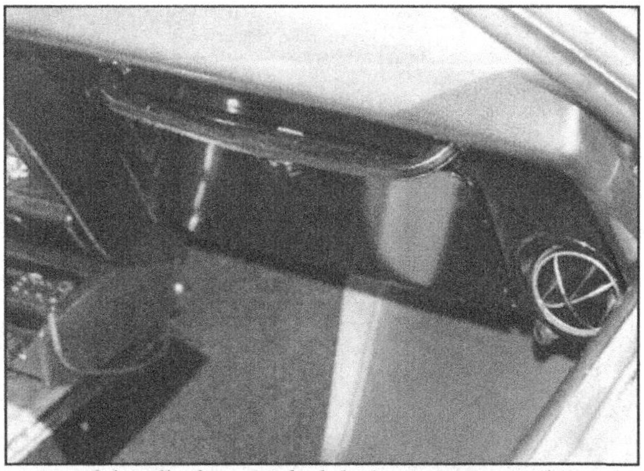

A grab handle that matched the instrument panel in color was used only with the deluxe interior trim.

1968-1969 Deluxe Door panels.

1968 Door Panel Part Numbers					
Interior Code and Color		Standard Door Trim		Custom Door Trim	
		Passenger	Driver	Passenger	Driver
250	Teal	7761004	7761005		
251	Gold	7761006	7761007		
252	Red	7761007	7761009		
259,269	Black	7761010	7761011		
255	Blue			7772120	7772121
256	Turquoise			7772122	7772123
257	Gold			7772124	7772125
258	Red			7772126	7772127
253,272	Black			7772128	7772129
260,275	White			7772130	7772131
261	Turquoise	7786534	7786534		
262, 273	White	7786563	7786564		
281	Saddle			8708630	8708631

1968 Kick Panels

Color	**Drivers**	**Passengers**
Without Air Conditioning		
Black	7664672	7664671
Red	7664674	7664673
Gold	7734649	7734648
Teal	8760588	8760589
Turquoise	7761822	7761821
Saddle	8712517	8712516
With Air Conditioning		
Black	7664684	7664685
Red	7664694	7664695
Gold	7734639	7734638
Teal	7783417	7783416
Turquoise	7761826	7761824
Saddle	8712519	8712518

1968 Kick Panels

Color	**Drivers**	**Passengers**
Without Air Conditioning		
Black	7664672	7664671
Red	7664674	7664673
Gold	7734648	7734649
Teal	7783414	7783415
Turquoise	7761821	7761822
Saddle	8712516	8712517
With Air Conditioning		
Black	7664685	7664684
Red	7664695	7664694
Gold	7734638	7734639
Teal	7783416	7783417
Turquoise	7761824	7761825
Saddle	8712518	8712519

1968 Arm Rest		
Color	**Drivers**	**Passengers**
Standard Trim		
Black	7766490	7766491
Red	7766496	7766497
Gold	7766494	7766495
Teal	7766498	7766499
White	7766806	7766807
Turquoise	7766802	7766803
Saddle	8803586	8803587

1969 Door panels with standard trim

1969

Visually in design the custom door panels were the same units used in 1968. The standard trim door panel featured two rods of vertical pleats in the Madrid grain vinyl that were separated by two thin chrome strips; a larger chrome molding was place on the top portion of the panel.

Door Moldings with Standard Trim

	W/O Remote Control Mirror	With Remote Control Mirror
Upper	Left-Hand 7737762 Right-Hand 7737762	Left-Hand 7726724-front 7759450-rear Right-Hand 7737762

1969 Arm Rest

Color	Drivers	Passengers
Standard Trim		
Black	7766490	7766491
Red	7766496	7766497
Gold	7766494	7766495
Teal	7766498	7766499
White	7766806	7766807
Green	7766802	7766803

1969 Door Panel Part Numbers

Interior Code and Color		Standard Door Trim		Custom Door Trim	
		Passenger	Driver	Passenger	Driver
200	Blue	8739498	8739499		
202	Gold	8739500	8739501		
206	Green	8766202	8766203		
207	Parchment	8739504	8739505		
208	Black	8739506	8739507		
210	Blue			8739508	8739509
212,293	Gold			8739510	8739511
214	Red			8739512	8739513
216	Green			8739514	8739515
217,227	Parchment			8739516	8739517
218,228	Black			8777504	8777505

Rear Quarter Panel Trim
1967-1969

The rear quarter panels match the door panels in design. However, while the same door panels were used on hardtop and convertible models the rear quarter panel is drastically different between the hardtop and the convertible models. Color choices were the same as they were on the door panels. Bright molding was used on the same axis as the door panel. Part number 7669270 was used the 1967 hardtop with custom trim, no trim molding was used on the convertible. In 1968 and 1969 the part number was changed to 7752625 (upper molding) for the hardtops and the convertibles used part number 7737770.

1969 Kick Panels

Color	Drivers	Passengers
Without Air Conditioning		
Black	7664672	7664671
Red	8760585	8760584
Gold	8760587	8760586
Blue	8760589	8760588
Green	8760591	8760590
With Air Conditioning		
Black	7664684	7664685
Red	8728243	8728242
Gold	8728245	8728244
Blue	8728247	8728246
Green	8728249	8728248

1967 Rear Arm Rest

Color	Drivers	Passengers
Standard/Custom Trim		
Black	7716754	7716755
Red	7716752	7716753
Gold	7716748	7716749
Blue	7716764	7716765
White	7716768	7732569
Bright Blue	7716756	7716757
Turquoise	7716766	7716767

1968 Rear Arm Rest

Color	Drivers	Passengers
Standard/Custom Trim		
Black	7782590	7782591
Red	7782596	7782597
Gold	7782594	7782595
Teal	7782758	7782759
White	7782756	7782757
Turquoise	7782760	7782761
Saddle	8765198	8765199

1969 RearArm Rest

Color	Drivers	Passengers
Standard/Custom Trim		
Black	7782590	7782591
Red	8765228	8765229
Gold	8765234	8765235
Blue	8765226	8765227
White	8765236	8765237
Green	8765232	8765233

1967 Rear Quarter Panel Part Numbers

Interior Code and Color		Standard Trim				Custom Trim			
		Passenger		Driver		Passenger		Driver	
		Hdtp.	Conv.	Hdtp.	Conv.	Hdtp.	Conv.	Hdtp.	Conv.
250, 270	Blue	7714046	7714048	7714047	7714049				
251, 271	Med. Gold	7673710	7646520	7673711	7646521				
252	Red	7673714	7646524	7673715	7646525				
253. 272	Black	7673716	7646526	7673717	7646527				
254	White	7728348	7732516	7728349	7732517				
255, 265	Bright Blue					7673724	7673738	7673725	7673739
256	Turquoise					7714052	7714060	7714053	7714061
257, 267	Gold					7673720	7673734	7673721	7673735
258	Red					7673726	7673740	7673727	7673741
259. 269	Black					7673728	7673742	7673729	7673743
260	White					7714054	7714062	7714054	7714063

1968 Rear Quarter Panel Part Numbers

Interior Code and Color		Standard Trim				Custom Trim			
		Passenger		Driver		Passenger		Driver	
		Hdtp.	Conv.	Hdtp.	Conv.	Hdtp.	Conv.	Hdtp.	Conv.
250	Teal	7769726	7769736	7769727	7769737				
251	Gold	7769728	7769738	7769729	7769739				
252	Red	7769730	7769740	7769731	7769741				
259,269	Black	7769732	7769742	7769733	7769743				
255	Blue					7781732	7781708	7781733	7781709
256	Turquoise					7781734	7781710	7781735	7781711
257	Gold					7781736	7781712	7781737	7781713
258	Red					7781738	7781714	7781739	7781715
253,272	Black					7781740	7781716	7781741	7781717
260,275	White					7781742	7781718	7781743	7781719
261	Turquoise	7786546	7786550	7786547	7786551				
262, 273	White	7786548	7786552	7786549	7786553				
281	Saddle					8708646	7798088	8708647	7798089

1969 Rear Quarter Panel Part Numbers

Interior Code and Color		Standard Trim Passenger Hdtp.	Standard Trim Passenger Conv.	Standard Trim Driver Hdtp.	Standard Trim Driver Conv.	Custom Trim Passenger Hdtp.	Custom Trim Passenger Conv.	Custom Trim Driver Hdtp.	Custom Trim Driver Conv.
200	Blue	8739668	8739812	8739669	8739813				
202	Gold	8739670	8739814	8739671	8739815				
206	Green	8766196	8766200	8766197	8766201				
207	Parchment	8739672	8739816	8739673	8739817				
208	Black	8739674	8739818	8739675	8739819				
210	Blue					8739802	8739820	8739803	8739821
212,293	Gold					8739804	8739822	8739805	8739823
214	Red					8739806	8739824	8739807	8739825
216	Green					8739808	8739826	8739808	8739827
217,227	Parchment					8739810	8739828	8739811	8739829
218,228	Black					8739812	7781716	8739813	7781717

Carpeting

The original carpet was a cut pile type that was made from 100 % nylon. They were press molded and were color-keyed to the interior trim; except for the white (parchment) trim which used a black colored carpet. A big misconception about the carpeting was that they all were done in a two-piece design (front and rear) but this is not true. Those models without a console were done in a two-piece design, but those with a console were done in a three-piece design: 2-front pieces and 1-rear piece. Also due to panel reinforcement in the floor pan of convertible models, the ragtop Firebird used a different carpet, even though they were the exact same color.

1967 Firebird Carpeting Part Numbers

Interior Trim Codes	Carpet Color	Hardtop Front	Hardtop Rear	Convertible Front	Convertible Rear
253. 272, 254,259.	Black	7657046*	7657058#	7657052*	
269, 260		7666912-R.H**	7666924##	766900-R.H**	
		7666913-L.H.**		766901-L.H. **	
252,258	Red	7657049*	7657061	7657055	
		7666918- R.H**			
		7666919- L.H**			
251,257,267,271	Gold	7657051			
	Blue				
	Turquoise	7728869*		7728854*	
		7728885-R.H.**		7728862-R.H.**	
		7728886 L.H.**		7728863-L.H.**	

*- Without console **- with Console # Bench seat ##- with Bucket seats

1968 Firebird Carpeting Part Numbers

Interior Trim Codes	Carpet Color	Hardtop Front	Hardtop Rear	Convertible Front	Convertible Rear
253. 272, 254,259. 269, 260	Black	7657046* 7666912-R.H** 7666913-L.H.**	7779018# 7779023##	7657052* 766900-R.H** 766901-L.H. **	7779029
252,258	Red	7657049* 7666918- R.H** 7666919- L.H**	7779024	7657055	7779030
250,255	Teal	7785101* 7773824-R.H. ** 7773825-L.H. **	7779026	7785102* 7785074 R.H.** 7785075 L.H.**	7779032
251,257	Gold	7728867* 7728883-R.H.** 7728884 L.H.**	7779028	7728852* 7728860-R.H.** 7728861-L.H.**	7779034
256,261	Turquoise	7728869* 7728885-R.H.** 7728886 L.H.**	7779027	7728854* 7728862-R.H.** 7728863-L.H.**	7779033
281	Saddle	7728870* 7728887- R.H.** 7728888 -L.H.**	7799672	7728855* 7728864-R.H.** 7728865-L.H.**	7799671

*- Without console **- with Console # Bench seat ##- with Bucket seats

1969 Firebird Carpeting Part Numbers

Interior Trim Codes	Carpet Color	Hardtop Front	Hardtop Rear	Convertible Front	Convertible Rear
208, 218, 228, 207, 217, 227	Black	7657046* 7666912-R.H** 7666913-L.H.**	7779018# 7779023##	7657052* 766900-R.H** 766901-L.H. **	7779029
214	Red	8733534* 8733542-R.H. ** 8733543-L.H. **	8733563	8733528* 8733550-R.H ** 8733551-L.H **	8733569
200,210,	Blue	8733524* 8733544-R.H ** 8733545-L.H **	8733558	8733527* 8733548 R.H.** 8733549 L.H.**	8733564
202,212, 293	Gold	8733530* 8733536-R.H. ** 8733537-L.H. **	8733562	8733544* 8733544-R.H.** 8733545-L.H.**	8733568
206,216	Green	8733532* 8733538-R.H. ** 8733539 -L.H. **	8733561	8733526* 8733546-R.H.** 8733547-L.H.**	8733567

*- Without console **- with Console # Bench seat ##- with Bucket seats

Headliner

Two different grains were used in 1967. Those with the standard trim used a Tier grain, while those with the custom trim used a Premier Perforated grain. In 1968 the models with standard trim used Bedford ribbed grain, and the Tier grain with standard trim, while those with custom trim again used the Premier Perforated headliners. In 1969 both the standard and custom trim used the Premier Perforated headliners. However, there have been cars where the Tier grain was used with Custom trim, or the Premier grain with standard trim. This usually occurs on cars with black or white trim, where a shortage accorded on the line and a substitute was used. Documenting this would be impossible, but we have seen at least two examples of this in unrestored Firebirds that were sitting in the salvage yard. The headliner matched the interior trim in color, and was used only on the hardtop models. No headliner was used in convertibles.

1967 Headliner Part Numbers

Code	Color	Part Number	Grain
250,270	Dark Blue	7713939	Tier
251, 271	Med. Gold	7713940	Tier
252	Red	7713941	Tier
253,272	Black	7713942	Tier
254	White	7732512	Tier
255,265	Bright Blue	7728350	Premier
256	Turquoise	7713944	Premier
257,267	Gold	7713945	Premier
258	Red	7713946	Premier
259,269	Black	7713947	Premier
260	White	7713948	Premier

1968 Headliner Part Numbers

Code	Color	Part Number	Grain
250	Teal	7771780	Tier
251	Gold	7776009	Tier
252	Red	7766010	Tier
259,269	Black	7713947	Tier
255	Blue	7771781	Premier
256	Turquoise	7771781	Premier
257	Gold	7771783	Premier
258	Red	7713946	Premier
253,272	Black	7766007	Premier
260,275	White	7771784	Premier
261	Turquoise	7786538	Premier
262, 273	White	7786539	Premier
281	Saddle	8708636	Premier

1969 Headliner Part Numbers

Code	Color	Part Number	Grain
200,210	Blue	8725445	Premier
202,212,293	Gold	8739484	Premier
206,216	Green	8725448	Premier
207,217,227	Parchment	8739485	Premier
208,218,228	Black	7766007	Premier
214	Red	8725446	Premier

1967 Sun Visor Part Numbers

		Hardtop		Convertible	
Code	Color	Part Number	Grain	Part Number	Grain
250,270	Dark Blue	7713950	Tier	7713973	Madrid
251, 271	Med. Gold	7713963	Tier	7666119	Madrid
252	Red	7713964	Tier	7666121	Madrid
253,272	Black	7713965	Tier	7666122	Madrid
254	White	7732513	Tier	7713975	Madrid
255,265	Bright Blue	7732504	Premier	7666124	Madrid
256	Turquoise	7713967	Premier	7713974	Madrid
257,267	Gold	7713968	Premier	7666119	Madrid
258	Red	7713969	Premier	7666121	Madrid
259,269	Black	7713970	Premier	7666122	Madrid
260	White	7713971	Premier	7713975	Madrid

1968 Sun Visor Part Numbers

		Hardtop		Convertible	
Code	Color	Part Number	Grain	Part Number	Grain
250,270	Teal	7769798	Tier	7769853	Madrid
251,271	Gold	7766103	Tier	7766107	Madrid
252	Red	7766104	Tier	7766108	Madrid
253,272	Black	7766101	Tier	7666122	Madrid
255,265	Blue	7769851	Premier	7769853	Madrid
256	Turquoise	7769800	Premier	7766109	Madrid
257	Gold	7769799	Premier	7766107	Madrid
258	Red	7766104	Premier	7766108	Madrid
259,269	Black	7713970	Premier	7666122	Madrid
260,275	White	7769852	Premier	7666122*	Madrid
261	Turquoise	7786544	Premier	7766109	Madrid
262, 273	White	7786545	Premier	7666122*	Madrid
281	Saddle	8708637	Premier	8708645	Madrid

1969 Sun Visor Part Numbers

		Hardtop		Convertible	
Code	Color	Part Number	Grain	Part Number	Grain
200,210	Blue	8765282	Premier	8765290	Madrid
202,212,293	Gold	8765286	Premier	8765294	Madrid
206,216	Green	8765285	Premier	8765293	Madrid
207,217,227	Parchment	8765287	Premier	8765289*	Madrid
208,218,228	Black	8765281	Premier	8765289	Madrid
214	Red	8765283	Premier	8765291	Madrid

Inside Mirror
1967-1968

A bright plated back mirror part number 3633013 was standard in all Firebird models. The mirror support should have a natural appearance and differed according to model year and body style.

Colored cap, that was color-keyed to the interior trim, except the 1967 convertible, which used gray, was used on the support. For 1967 the Firebird used support caps that were unique to this model line only. For 1968 the cap on the hardtops is the same unit used in the Tempest line, but the convertible continued to use special parts.

1967-1969 Mirror Support

1967	Convertible	9711724
1967	Hardtop	7729476
1968-1969	Hardtop	8752692
1968-1969	Convertible	7740300

1967Mirror Support Caps

Hardtop		Convertible	
Color	Part No.	Color	Part No.
Black	7729515	Gray	9711985
Red	7729517		
Gold	7729520		
Dark Blue	7729521		
Bright Blue	7729523		
Parchment	7729518		
Aqua	7729519		

1968 Mirror Support Caps

Hardtop Color	Part No.	Convertible Color	Part No.
Black	7764898	Black	7767945
Red	7764899	Red	7767946
Gold	7764904	Turquoise	7767947
Teal	7764902	Teal	7767949
Saddle	7764903	Gold	7767950
Parchment	7764905	Saddle	8712527
Turquoise	7764900		

1969

The mirror was completely restyled for 1969 and will not fit earlier models. The mirror was color-keyed to the interior trim, except for white, which used a black colored mirror. The support was the same unit used on the 1968, while the 1968 mirror will physically fit, it is not correct for a 1969 model. Color-keyed caps were again used on the supports.

1969 Mirror Part Numbers

Color	Part No.	Color	Part No.
Black	911250	Blue	911251
Green	911252	Red	911253
Gold	911254		

1969 Mirror Support Caps

Hardtop Color	Part No.	Convertible Color	Part No.
Black	7764898	Red	8751958
Red	8752008	Gold	8751959
Gold	8752009	Blue	8751960
Blue	8752011	Green	8751962
Green	8752014	Black	7767945
Parchment	8752007		

1968-1969 Console Front Trim Panels

Transmission.	Without Clock	With Clock
1968		
Manual	9793253	9793254
Automatic	9793129	9793252
1969		
Manual	9793253	
Automatic	9793129	

Console
1967-1969

The console was an option on all Firebirds, however, it was not available if the bench seat was ordered. The base was listed as part number 3918725 in 1967 and 9793278 in 1968-69 and was painted to match the interior trim, except for white (parchment) trim, which was painted black. The finish is a 60 degree gloss in 1967 and 30-degrees in 1968 and 1969.

The base held to the floor with a bracket part number 3893889, which was painted the same color as the lower portion of the car as this bracket was in place when the body was painted.

An escutcheon panel that trims the shift lever opening was used in 1968 and 1969 according to transmission type. All manual transmission used part number 9793260. All automatics used part number 9793256 in 1968, while in 1969 part number 9797443 was used on two-speed automatic and part number 9797444 was used on turbo-hydra matics. For 1968 a front trim plate was used which depended on the transmission type and whether or not the clock was ordered. When an optional clock was ordered it was mounted on the front upward face of the console and required a special trim plate. In 1969 the clock was moved to the instrument cluster and only the transmission type affected the trim plates usage. The console door was listed as part number 3918723 in 1967, and 9793229 in 1968 and 1969. Both doors used a 60-degree gloss black hinge listed as part number 3893866.

Seat and Shoulder Straps
1967-1969

Lap belts were standard on all Firebird models, shoulder straps became required equipment on January 1, 1969, before that date they were optional equipment. Seat belts came in two different styles. The standard belt, used a black colored buckle, and the custom belts used brushed metal buckles. For the most part the custom and the standard belts used colored webbing that matched the seat trim except for the parchment trim, which used black colored belts.

Note there were early and late style belts on the 1967 models. The early models have a Fisher coach symbol, while the later models have a mark of excellence logo, there is no date of this change noted, but the GM logo should not appear on early built Firebirds.

A rare option is the rear seat shoulder belts. They too came in colors. It was available only for 1968 and 1969, and only with standard belts.

1967 Lap and Shoulder Belt Part Numbers-Standard

Interior Color	Front Lap Belts		Front Shoulder Belts		Rear Seat Belts	
	Driver	Passenger	Driver	Passenger	Lap	Shoulder
Black/Parchment	8700619*	8700618*	7792095	7792095	8700461*	Not
	8700629**	8700628**	7792108(1)	7792108(1)	8700647**	Available
Turquoise	8700625*	8700626*	7792098	7792098	8700464*	
	8700634**	8700634**	7792111(1)	7792111(1)	8700650	
Blue	8700685*	8700686*	7792096	7792096	8700462*	
	8700631**	8700630**	7792109(1)	7792109(1)	8700648**	
Gold	8700623*	8700622*	7792097	7792097	8700463*	
	8700633**	8700632**	7792110(1)	7792110(1)	8700649**	

*_early style belt **-Late style Belt (1) convertible only

1967 Lap and Shoulder Belt Part Numbers-Deluxe

Interior Color	Front Lap Belts		Front Shoulder Belts		Rear Seat Belts	
	Driver	Passenger	Driver	Passenger	Lap	Shoulder
Black/Parchment	8700607	8700606	7792076	7792076	8700510	Not
			7992099(1)	7992099(1)		Available
Turquoise	8700613	8700612	7792082	7792082	8700517	
			7992105(1)	7992105(1)		
Blue	8700611	8700610	7792081	7792081	8700516	
			7992104(1)	7992104(1)		
Gold	8700615	8700614	7792083	7792083	8700518	
			7992106(1)	7992106(1)		
Red	8700609	8700608	7792080	7792080	8700512	
			7992100(1)	7992100(1)		
Bright Blue	8700617	8700616	7792143	7792143	8700519	
			7792107 91)	7792107 91)		

(1)- Convertible only

1968 Lap and Shoulder Belt Part Numbers-Standard

Interior Color	Front Lap Belts		Front Shoulder Belts		Rear Seat Belts	
	Driver	Passenger	Driver	Passenger	Lap	Shoulder
Black/Parchment	8701661	8701660	8700192 8701677(1)	8700192 8701677(1)	8701644	8701677
Teal	8700689	8700688	8700195 7794965(1)	8700195 7794965(1)	8700709	7794965
Gold	8700687	8700686	8700194 8700273(1)	8700194 8700273(1)	8700708	8700273

(1) Convertible only

1968 Lap and Shoulder Belt Part Numbers-Deluxe

Interior Color	Front Lap Belts		Front Shoulder Belts		Rear Seat Belts	
	Driver	Passenger	Driver	Passenger	Lap	Shoulder
Black/Parchment	8700691	8700690	8700196 8700264(1)	8700196 8700264(1)	8701501	Not Available
Red	8700693	8700692	8700197 7794976(1)	8700197 7794976(1)	8700747	
Teal	8700697	8700696	8700200 7794978(1)	8700200 7794978(1)	8700750	
Turquoise	8700699	8700698	8700201 8700268(1)	8700201 8700268(1)	8700751	
Gold	8700701	8700700	8700202 8700269(1)	8700202 8700269(1)	8700752	
Saddle	8700851	8700850	8700203 8700270(1)	8700203 8700270(1)	8700753	

(1) convertible only

1969 Lap and Shoulder Belt Part Numbers-Standard

Interior Color	Front Lap Belts		Front Shoulder Belts		Rear Seat Belts	
	Driver	Passenger	Driver	Passenger	Lap	Shoulder
Black/Parchment	8701661	8701660	8701690 8701677(1)	8701690 8701677(1)	8701644	Not Available
Blue	8701663	8701662	8701691 8701678(1)	8701691 8701678(1)	8701645	
Gold	8701674	8701673	8701692 8701679(1)	8701692 8701679(1)	8701646	

(1) Convertible only

1969 Lap and Shoulder Belt Part Numbers-Deluxe

Interior Color	Front Lap Belts		Front Shoulder Belts		Rear Seat Belts	
	Driver	Passenger	Driver	Passenger	Lap	Shoulder
Black/Parchment	8701592	8701591	8701599 8701526(1)	8701599 8701526(1)	8701501	Not Available
Blue	8701596	8701595	8701605 8701527(1)	8701605 8701527(1)	8701503	
Gold	8701598	8701597	8701606 8701528(1)	8701606 8701528(1)	870504	
Green	8701603	8701602	8701610 8701530(1)	8701610 8701530(1)	8701508	
Red	8701594	8701593	8701608 8701531(1)	8701608 8701531(1)	8701506	

(1) convertible only

Chapter 10

Electrical Systems

Starting and Ignition

1967

A Delco-Ramey top post 12-volt battery was standard in all Firebird models. A Y44 type battery was standard in the six cylinder 1-bbl models, unless air conditioning was ordered, whre the heavy-duty type R-59 rated at 61-amps was used. The heavy-duty battery was standard in all 4-bbl equipped model, sand was optional as a heavy duty unit for the 230-ci six cylinder and the 2-bbl 326-ci V-8. Standard battery in the 326-ci 2-bbl V-8 was a type Y59 that was rated at 53-amps, the heavy duty unit was used if air conditioning was ordered.

All six cylinder Firebirds used two different battery trays, part number 3893882 was used on models without air conditioning and part number 9789148 was used with air conditioning. All V-8 Firebirds used part number 3893882 regardless of whether or not air conditioning was ordered. The battery tray should be painted semi gloss black, and given a coating that will prevent it from corroding, this coating was not originally used, but will prevent the tray from disappearing over the years, (ever try to find a used battery tray?)

A black colored twin lead positive cable was routed from the battery to the starter motor. Cable part number depends on engine size

1967-68 Positive Battery Cable Part Numbers

Application	Part Number
6-cyld.	6290588*
	6289146**
V-8	6289140

-Std Battery **-H.D. Battery

The starter motor for six cylinders is listed as part number 1108329. Those with the 326-ci 2-bbl V-8 used part number 1108328, and all 4-bbl equipped V-8's used part number 1108355. The housing should be painted semi-gloss black, the nose should have a natural aluminum appearance.

The distributor was keyed to the engine size and output. Also those models with California emissions required special distributors than those used in the other states. The identification number is stamped on the distributor housing or a band around the unit.

1967 Distributor Identification Numbers

CID	Carb	Emissions	ID Number
230	1-bbl	No	1110396
230	1-bbl	Yes	1110691
230	4-bbl	No	1110397
230	4-bbl	Yes	1110398
326	2-bbl	No	1111164
326	2-bbl	Yes	1111199
326	4-bbl	No	1111165
326	4-bbl	Yes	1111238
400	4-bbl	No	1111250
400	4-bbl	Yes	1111252

Original batteries will have the Delco name imprinted into it.

Packard Electric supplied most of the spark plug cables and they will have the Packard logo on them. These cables were also date coded when they were made at the Packard Electric factory, before being shipped for installation at the Lordstown assembly plant. Because of this the date on the cables will be for the assembly date of your car. However, the date on these cables is different than other parts on your car. Much like the date on the hose clamps, this will have a code that will look similar to this 1Q67. The first digit represents the first quarter of the year. The last two are the year of manufacture. These cables would for 1967 models while the code 3Q67 would be for the third quarter of the 1967 model years and would be for early built 1968 models.

Three different plug sets were used. All six cylinders used part number 2986606, while the V-8's used two different sets. Those V-8 models without air conditioning used part number 6293790 while those with air conditioning used part number 6289616. Those with air conditioning required longer plug wires as they had to be route through an extra wire bracket.

1968

In 1967 an octagon head shaped key was used for the ignition. For 1968 a hex-head type ignition was used type "C" key was used. The ignition switch was listed as part number 116694 and was located in the instrument panel. Battery, battery trays, and cables used the same part numbers they used in 1967.

The starter motor assembly for six cylinders was listed as part numbers 11089329, those with the 350-ci 2-bbl used part number 1108328, while the 4-bbl V-8's used part number 1108446. Painting instructions for the starter were the same as the year before.

With the deletion of the A.I.R. emissions package this option no longer affected the distributor, however transmission type now affected their usage. Horsepower also affected the usage as Ram Air models required a special carburetor.

1968 Distributor Identification Numbers

CID	Carb	Transmission	ID Number
250	1-bbl	All	1110430
250	4-bbl	Manual	1110431
250	4-bbl	Automatic	1110449
350	2-bbl	Manual	1111281
350	2-bbl	Automatic	1111165
350	4-bbl	Manual	1111447
350	4-bbl	Automatic	1111282
400*	4-bbl	Manual	1111449
400	4-bbl	Automatic	1111270
400	4-bbl	RAM AIR II	1111941

*-Except Ram Air II

1969

The ignition switch was moved to the steering column and when in the lock position would lock the steering column in place. Two different switches were used. Part number 199086 was used without tilt steering and part number 199093 was used with tilt steering. The switches were used in all 1969 Pontiacs and will fit your Firebird. When swapping switches the cylinder can be recoded to fit your present key, any professional locksmith can do this. With the change in ignition switch position the key was also restyled. The head was now a square and was coded with the letter "E". Battery, trays remained the same. However, the battery cables changed.

1969 Positive Battery Cable Part Numbers

Application	Part Number
6-cyld.	8907356
V-8	6296734
1969 Negative Battery Cable	
All	2978482

1969 Distributor Identification Numbers

CID	Carb	Transmission	ID Number
250	1-bbl	All	1110475
250	4-bbl	All	1110474
350	2-bbl	Manual	1111960
350	2-bbl	Automatic	1111942
350	4-bbl	Manual	1111966
350	4-bbl	Automatic	1111965
400*	4-bbl	Manual	1111952
400*	4-bbl	Automatic	1111946
400	4-bbl	RAM AIR IV	1111941

*-Except Ram Air IV

1967

A Delco-Ramey built alternator was used in all applications. For the most part a 37-amp alternator was standard with all powerplants, and a 55-amp alternator was used if air conditioning or the heavy-duty alternator was ordered. However, even though the rating is the same, many different alternators were used according to engine displacement and options like A.I. R. emissions package that was mandatory in California, and in some cases power steering would affect the usage of an alternator.

1967 Alternators

Amp Rating	ID Number	Notes
	Six Cylinder	
37	1100761	W/O a/c or p/s
37	1100781	W/O a/c with p/s
55	1100760	a/c w/o A.I.R or p/s
55	1100782	With a/c and A.I.R.
55	1100745	With a/c and p/s
	V-8	
37	1100704	W/O a/c or p/s
37	1100768	W/O a/c or A.I.R.
55	1100770	With a/c
55	1100700	With a/c and p/s

Regulator was listed as part number 1119515 for all models and applications. Brackets used to hold the alternator in position depended on the engine size and steering type; California bound cars also required special brackets because of the Air Injector Pumps.

1967-1969 Alternator Brackets and Braces

Model Year	Part Number	Notes
	Six Cylinder	
1967-69	9782681	W/O A.I.R. or P/s
1967-69	9788044	With P/S
1967	9782682	With A.I.R.
	V-8	
1967	9774773	W/O/ A.I.R. or p/s
1967	9786904	With A.I.R. W/O p/s
1967-69	9786911	With p/s

1967 Engine Wiring Harness

Application	Part Number
ALL	
Man Trans. W/o console or rally cluster	6290648
Man. Trans with console w/o Rally cluster	6291285
Man trans. W/O console with Rally cluster	6293183
Man. trans. With console with rally cluster	6293181
Auto. Trans. W/o console or rally cluster	6290159
Auto. Trans with console w/o Rally cluster	6290650
Auto. trans. W/O console with Rally cluster	6293180
Auto. trans. With console with rally cluster	6293182

1968

Not much change; the rating for the alternators remained the same, but there was a revision of the alternators due to the deletion of the A.I.R. emissions.

1968 Alternators

Amp Rating	ID Number	Notes
	Six Cylinder	
37	1100761	W/O a/c or p/s
37	1100781	W/O a/c with p/s
55	1100760	With a/c w/o p/s
55	1100745	With a/c and p/s
	V-8	
37	1100704	W/O a/c or p/s
55	1100700	With a/c and p/s

1968 Engine Wiring Harness

Application	Part Number
ALL	
Man Trans. W/o rally cluster	6295747
Man trans. with Rally cluster	6295556
Auto. Trans. W/o console or rally cluster	6295747
Auto. Trans with console w/o Rally cluster	6295749
Auto. Trans. W/O console with Rally cluster	6295554
Auto. Trans. With console with rally cluster	6295556

1969

Rating for the alternators remained the same, but were again juggled with different identification numbers and those models with a 400-ci power plant required a special alternator.

1969 Alternators

Amp Rating	ID Number	Notes
	Six Cylinder	
37	1100761	W/O a/c or p/s
37	1100781	W/O a/c with p/s
55	1100760	With a/c w/o p/s
55	1100745	With a/c and p/s
	350- V-8	
37	1100704	W/O a/c or p/s
55	1100700	With a/c and p/s
	400-ci V-8	
37	1100832	W/O a/c
55	1100830	With a/c
All models with heated back glass option		
55	1100854	All

1969 Engine Wiring Harness

Application	Part Number
ALL	
Except Rally Cluster or Aux. Gauges	6297444
Aux Gauges	6299356
With Rally Gauges	6297446

Exterior Lighting
1967

The headlamp switch was located to the left of the driver, it was listed as part number 1995138 and was the same unit used in 1964-1967 Tempests. The bright plated knob part number 9783323 was also used on all 1966-1968 Pontiac models. The bezel with the lettering on it was listed as part number 3923943. However, some Firebirds used part number 9786945, which can be found on the full-size Pontiacs. The switch controlled the quad headlamp arrangement and tail lamps along with the instrument cluster lamp lever.

The headlamps were listed as part numbers 5959713 left-hand and 5959714 right-hand. The tail lamps consisted of a design of two thin horizontal strips that were parallel to each other and were mounted in the rear end panel of the car. However, the tail lamps on each side were one unit. Styling slots gave the appearance of a quad tail lamp arrangement.

Tail lamp housings were listed as part number 5959712, and could be used on either side. Tail lamp lenses were red in color and were listed as part number 5959716, like the housings they fit either side of the car. The screws used to mount the tail lamp lenses were the same units used to mount the lens on the 1966-1967 Grand Prix.

Signal lever was mounted on the left-hand side of the steering column two different levers were used. Those Firebirds without cruise control used part number 977899, while those with cruise control used part number 6465211, which supported the cruise control controls. The front turn lamps were positioned under the front bumper and were listed as part numbers 916533 left-hand and 916534 right-hand.

1967 Firebird used twin round shaped headlamps on each side. The front turn lamps were placed below the front bumper.

1968 Firebird front turn lamps wrapped around the bottom edge of the front pan and front fender to double as side running lamps.

1968

The headlamp switch was listed as part number 1995154 and was the same unit that was used on the 1968-1971 Tempest models. The knob used the same part number that it used the year before. The bezel was listed as part number 392343.

Headlamp arrangement was the same as in 1967, and both they and the tail lamps used the same parts as the year before. Front parking had to be restyled due to federal regulation of side running lamps. Instead of altering the fender design Pontiac engineers styled the front turn lamp so that It would wrap around and double as a front side lamp. The front turn lamps were listed as part numbers 5960531 Left-hand and 5960532 left-hand.

1969

Headlamp switch used the same part number as it did in 1968, but the knob was switched to part number 979087 and the bezel was listed as part number 9798659. Headlamps remained a quad arrangement but were restyled and were broken down into inner and outer parts. Note: as in this case and the others before, we're discussing the headlamp assembly as a whole, not just the headlamp element.

1969 Headlamp Assembly-Less Door

	Inner	Outer
Left-hand Side	5963838	5963839
Right-hand Side	5963837	5963840

Tail lamps had the same general twin horizontal slot design as before, but were now more pronounced than in previous years. The housings were also unique to each side of the car. Listed as part numbers 5962109 left-hand and 5962110- right-hand they used a lens listed as part number 5961294. Unlike the housings, the lens will fit either side of the car.

Front turn lamps were also restyled to give the car a fresher newer look and tie in with the other Pontiacs. Listed as part number 910363, it featured a bezel that looked like the cross hairs of a rifle. Unlike those in 1968, these front lamps did not wrap around the fenders thus an additional lamp part number 911138 had to be placed in the fenders, this is the same unit that was used on the 1969 Tempest models. The rear marker doubled as a medallion and rear side marker lamp, it was uniquely styled like the Firebird medallions part number 91139, and was used only on the 1969 Firebird. A bezel part number 976219 completed the look.

The rear tail laps were similar in design to the 1967-68 models but the housings were designed to fit a specific side of the car.

1969 Front turn lamps

The 1969 Front fender running lamp was a small round unit placed down low on the car.

1968 Rear quarter panel marker was shaped like the Pontiac arrowhead.

1969 Rear side running lamp is used only on the Firebird models.

Firebird Engine Wiring Harness Part Numbers		
Engine Type	**Part Number**	**Notes**
	1967	
6-cyld.	6291812	Manual Transmission
6-cyld.	6291866	Automatic Transmission
V-8	6291813	Manual Transmission
V-8	6291867	Automatic Transmission
	1968	
6-cyld.	6295564	Manual Transmission
6-cyld.	6295566	Automatic Transmission
V-8	6295565	Manual Transmission
V-8	6295567	Automatic Transmission
	1969	
6-cyld.	6299276	Without a/c or 55-amp alt.
6-cyld.	6297463	With a/c without 55-amp C.S.I. alt.
6-cyld	6297423	With 55-amp C.S.I. Alt.
V-8	6297465	Without 55-am C.S>I. Alt.
V-8	6297568	With 55-amp C.S. I. Alt.
C.S.I. Alternator with internal regulator		

Wiper Systems 1967-1969

A two-speed electric windshield wiper motor part number 4911476 was standard in all 1967 Firebirds. The motor housing should be painted under hood black, which is a semi-gloss finish. The switch, part number 1993372, was located in the instrument panel to the right of the driver. The knob part number 9786797, was used in all 1967 Pontiac models, but the switch was used only in Tempest and Firebird models. The six-inch high opaque white washer jar was listed as part number 3840083, and was used all three years. It was held in a semi-gloss black bracket part number 548938. The under hood nozzle should have a natural appearance, it was listed as part number 9789340 and was used on both sides of the hood.

The motor in 1968 was listed as part number 4918442 that was used again in 1969. The motor housing was finished in the same manner as before. The switch assembly used the same part number as the year before but the knob was changed to part number 9790960, which again was used in all 1968 Pontiac models. The bezel ,part number 3923941, was used in both the 1967 and 1968 Firebirds.

While the washer jar and bracket were the same there is an important change that occurred in the washer nozzles. On the passenger's side of the hood part number 9789340 was used all year long. However, on the driver's side, early models used part number 9789340, which was six inches in length. Later models used a smaller square shaped unit that was 2-1/2x2-1/2 in. in length. There is no date known for this change, but it would be incorrect for an early built Firebird to use the smaller nozzle.

The same motor that was used in 1968 was used again in 1969. The switch was changed to part number 1993453, and included the knob. The washer jar was the same unit as the previous two years. The outlet nozzle on the driver's side was the same unit that was used on later built 1968 Firebirds; but the passenger's side was changed to part number 9797525.

The wiper arms were the same throughout the three-year run, but they are body style specific. Part number 3916690 was used on either side on the hardtop, and part numbers 3914751 left-hand and 3914752 right-hand were used on the convertibles. Either an Anco or Trico blade insert was used.

Heater Systems
1967-1969

The heater motor assembly part number 4918686 for all three years, this is the standard heater motor. Those with air conditioning used a different motor.For these part numbers see the Options chapter of this guide. The motor assembly is attached to a blower case, part number 3016905, which is located on the right-hand side of the firewall all three years. This case was used on all 1966-1970 Pontiac models without air conditioning. The heater core was listed as part number 3013319 in 1967 and 1968 and part number 3016842 in 1969.

The heater controls were located in the center of the instrument panel. For 1967 and 1968 the part number was listed as part number 7303092, for 1969 it was changed to part number 7307802. Hot air was directed through ductwork. The center/main air distribution duct was regulated by the fact of whether a tape player or console was ordered.

Chapter 11

Regular Production And Dealer Installed Options

Both Regular Production (those installed at the assembly plant) and Dealer Installed options (installed solely by the dealer) were available all three years for the Firebird. The 1967 and 1968 models were coded by dealer sales code. This three-digit number was the code that was used on the dealers order form. Beginning in 1969 these codes were replaced with regular production option codes. These code begin with a letter, which indicates the option group, (for example the letter "L" indicates the engine option group) the additional code represents that individual option.

In this section listed first are the options that were available each year. Those in **bold** are covered in detail in this chapter. Note that some options like power brakes or power steering are covered elsewhere in this guide under that particular heading. For example the option power steering is covered under the heading steering systems. If that option is covered in detail elsewhere than it is in *italics.*

Dealer installed options may some time mirror the factory installed option, such as outside mirrors. The difference between these options was, (besides being installed by the dealer), that dealer installed options are coded by a part number, instead of a sales or RPO code. For a dealer-installed option to be considered correct there must be paperwork that backs up the claim and that the dealer for the original owner installed it, mostly before they (the original owner) took delivery of it.

1967 Firebird Dealer Installed Options	
Sales Number	**Option**
948883	Tail Pipe Extension (Dual Exhaust V-8)
984114	Rear Floor Amts Ivory
984528	Wire Wheel Disc
984642	750 Watt Engine Block Heater
984644	400 Watt Engine Block Heater
984701	Locking Spare Tire
984819	Stereo Speaker Package
984825	Locking Gas Cap
984826	Trunk Floor Mat
984852	Floor Mat Dark Blue
984854	Floor Mat Light Red
984855	Rear Floor Mats Black
984859	Rear Floor Amts Red
984861	Floor Amt Ivory
984867	AM/FM Multiplex Radio Package
984868	Tissue Dispenser
984872	Power Drum Brakes
984877	Pushbutton Am Radio
984882	Tail Pipe Extension (6-cylider)
984887	Ram Air Inlet Package
984893	Floor Mat Light Blue
984894	Rear Floor Amts Light Blue
984895	Power Disc Brakes
984898	AM/FM Multiplex Radio Package Bracket with a/c
984938	Power Steering Adapter Package 6-cyld.
984989	Walnut gearshift Knob (manual transmission only)
985592	Fire Extinguisher
986602	Litter Container Blue
986603	Litter Container Fawn
986604	Litter Container Turquoise
986607	Litter Container Black
986608	LitterContainer Red
988395	Power Steering Adapter Package V-8
988544	Stereo Switch Package
988545	Front Speakers
988572	Rear Seat Speaker
988596	Outside mirror (passenger side)
988826	Compass
993537	Emergency Road Package
993970	Child Safety Seat
994186	Heater Unit (lower Radiator Hose)

1967 Firebird Factory Installed Options

Sales Code	Option
321	*Foam cushion (Front Seat)*
331	*Mirror group*
332	*Lamp Group*
341	Manual Rear Antenna
342	Pushbutton AM Radio
342	*230-ci Sprint Engine*
344	Pushbutton AM/FM radio
347	*400-c Ram Air V-8*
348	*400-ci 4-bbl V-8*
351	Rear Seat Speaker
351	*T.H. Automatic*
352	*Automatic Transmission*
354	Delco Stereo Tape Player
361	Heavy-Duty Air Cleaner
374	**Rear Window Defogger**
382	Door Edge Guards
391	Vanity Mirror
394	Remote-Control OSRV Mirror
401	**Trunk Lamp**
402	Ignition Switch Lamp
421	**Underhood lamp**
431	*Custom Seat Belts*
434	*Front Shoulder Belts*
441	*Cruise Control*
442	*Safeguard Speedometer*
444	*Rally Gauges*
452	*Wire Wheel Disc*
452	*Wirte Wheel Covers*
453	*Rally II wheels*
454	*Rally I wheels*
458	*Custom Wheel Covers*
461	*Deluxe Wheel Covers*
462	*Deluxe Steering Wheel*
471	*Custom Sport Steering Wheel*
472	*Console (Not Available with Bench Seat)*
474	*Clock*
481	*Dual Exhaust (350-2-bbl)*
491	*Rally Stripes*
494	*Dual Horns*

Sales Code	Option
501	*Power Steering*
502	*Power Brakes*
504	Tilt Steering Column
514	*Heavy-Duty Fan 326*
521	*Front Disc Brakes*
524	Custom Shift Knob
531	*Tinted Glass, All Glass*
544	*Power Convertible Top*
551	**Power Windows**
572	Headrests
582	**Air Conditioning**
621	*Ride and Handling package*
631	*Front Floor mats*
632	*Rear Floor Mats*
654	*Fold Dow Rear Seat Back*
674	*Heavy-Duty Alternator*
678	*Heavy Duty Alternator*
681	*Heavy-Duty Radiator*
684	*Heavy-Duty Fan Sprint*
704	*Hood Tachometer*
704	**Hood Tachometer**
731	*Safe-T-Track Rear Axle*
738	*Rear Axle Options*

1968 Firebird Dealer Installed Options	
Sales Number	**Option**
948883	Tail Pipe Extension (Dual Exhaust V-8)
984114	Rear Floor Amts Ivory
984528	Wire Wheel Disc
984642	750 Watt Engine Block Heater
984644	400 Watt Engine Block Heater
984701	Locking Spare Tire
984825	Locking Gas Cap
984826	Trunk Floor Mat
984848	Walnut gearshift Knob (manual transmission only)
984852	Floor Mat Dark Blue
984854	Floor Mat Light Red
984855	Rear Floor Mats Black
984859	Rear Floor Amts Red
984861	Floor Amt Ivory
984882	Tail Pipe Extension (6-cylider)
984898	AM/FM Multiplex Radio Package Bracket with a/c
984905	Tissue Dispenser
984938	Power Steering Adapter Package 6-cyld.
984963	Power Steering Package 6-cylder
984971	Cruise Control
984995	Pushbutton Am Radio
984997	AM/FM Multiplex Radio Package
985592	Fire Extinguisher
986602	Litter Container Blue
986603	Litter Container Fawn
986604	Litter Container Turquoise
986607	Litter Container Black
986608	LitterContainer Red
988395	Power Steering Adapter Package V-8
988544	Stereo Switch Package
988545	Front Speakers
988547	Rear floor Mat Turquoise
988549	Rear Floor Mat Light Blue
988550	Floor Mat Turquoise
988551	Floor Mat Gold
988552	Floor Mat Light Blue
988568	Hood Mounted Tachometer-6cyld.
988572	Rear Seat Speaker
988596	Outside mirror (passenger side)
988664	Tire Pump
988826	Compass
993537	Emergency Road Package
993970	Child Safety Seat
994186	Heater Unit (lower Radiator Hose)

1969 Firebird Dealer Installed Options	
Sales Number	**Option**
984644	400 Watt Engine Block Heater
984642	750 Watt Engine Block Heater
988587	AM FM Radio
988696	Arctic Wiper Blade
993970	Child Safety Seat
988600	Clock
988826	Compass
988651	Cruise Control
988555	Door Edge Guard
993537	Emergency Road Package
988723	Fender Splash Guard
985592	Fire Extinguisher
984861	Floor Mat Ivory
988622	Floor Mat Dark Blue
988624	Floor Mat Gold
988621	Floor Mat Red
988623	Floor Mats Green
988700	Front License Plate Frame
988683	Front Speakers
994186	Heater Unit (lower Radiator Hose)
988595	Hood Hold Down Pins
988663	Hood Mounted Tachometer-6cyld.
986607	Litter Container Black
986602	Litter Container Blue
986603	Litter Container Fawn
986608	Litter Container Red
984825	Locking Gas Cap
988596	Outside mirror (passenger side)
988669	Power Brake Package
988672	Power Steering Adapter Package V-8
984963	Power Steering Package 6-cylder
988715	Powerflo Ventilation Sysytem
988592	Pushbutton Am Radio
988581	Rear Antenna
993645	Rear Deck Luggage Carrier
984114	Rear Floor Amts Ivory
988630	Rear Floor Mats Dark Blue
988632	Rear Floor Mats Gold
988631	Rear Floor Mats Green
988629	Rear Floor Mats Red
988699	Rear License Plate Frame
988590	Rear Seat Speaker
988708	Rear Spoiler
988642	Remote Control OSRV Mirror
988707	Spare Wheel Cover
988609	Deluxe Wheel Cover
988664	Tire Pump
984905	Tissue Dispenser
984826	Trunk Floor Mat
993551	Trunk Lamp
988661	Under Hood Lamp
988825	Vanity Mirror
984528	Walnut gearshift Knob
984528	Wire Wheel Disc
988693	Remote Deck Lid Release
988611	Stereo Tape Player
988612	Stereo Tape Player with Console
988607	Custom Wheel Cover

1968 Firebird Factory Installed Options

Sales Code	Option
321	*Front Foam Cushion*
322	*Rear Window Defogger/Protection Group*
331	*Mirror Group*
332	*Lamp Group*
342	*250-ci 4-bbl*
343	*350-2bbl V-8*
344	*350-ci H.O. 4-bbl*
345	*400-ci 4-bbl*
347	*400-ci 4-bbl Ram Air*
348	*400-ci H.O.*
351	*T.H. 400 Automatic Transmission*
352	*2-speed Automatic*
354	*4-speed Transmission*
355	*Heavy-Duty 3-speed Floor Shift*
358	*Close Ratio 4-speed*
361	*Safe-T-Track Rear Axle*
361	*Heavy-Duty Safe-T-Track Axle*
381	**Manual Rear Antenna**
382	**Push-button AM Radio**
384	**Pushbutton AM/FM Radio**
391	**Rear Seat Speaker**
394	**Stereo Tape Player**
402	*Spare Wheel Cover*
404	**Rear Window Defogger**
412	Door Edge Guards
414	Dual Horns
421	Vanity Mirror (R.H.)
422	Vanity Mirror (L.H./R.H.)
424	*L.H. Remote Control OSRV mirror*
431	*Front And Rear Custom Seat Belts*
432	*Rear Shoulder Belts*
434	*Hood Mounted Tachometer*
441	**Cruise Control**
442	*Safeguard Speedometer*
444	*Auxiliary Gauge Cluster*
452	*Wire Wheel Covers*
453	*Rally II Wheels*
458	*Custom Wheel Covers*
461	*Deluxe Wheel Covers*
462	*Deluxe Steering Wheel*

Sales Code	Option
471	*Sport Steering Wheel*
472	*Console N/A with bench seat*
474	*Electric Clock N/A with Tape Player*
481	*Dual Exhaust*
482	*Exhaust Extensions*
492	Remote-Control Deck Lid
494	*Rally Side Stripes*
501	*Power Steering*
502	*Power Brakes*
504	*Tilt-Steering Column*
514	*Heavy Duty Fan 7-blade Fan*
521	*Front Disc Brakes*
524	*Custom Gearshift Knob (Man. Floor only)*
531	*Tinted Glass, All*
532	*Tinted Windshield Only*
534	*Custom Pedal Trim*
544	*Power Convertible Top*
551	**Power Windows**
554	*Custom Trim*
561	Power 4-Way Bench Seat
564	**Power 4-Way Bucket Seat**
568	*Bench Seat*
571	*Headrests*
582	**Custom Air Conditioning**
591	Speedometer Gear Adapter
604	*Rear Folding Seat Back*
621	*Ride and Handling Package*
631	*Front Mats*
632	*Rear Floor Mats*
634	*Adjustable Front and Rear Shock Absorbers*
652	Luggage Lamp
671	Under Hood Lamp
672	Ignition Switch Lamp
701	*Heavy-duty Battery*
731	*Heavy-Duty Air Cleaner*
732	*Trunk Mat*
STT	*Special Two-Tone*
SVT	*Vinyl Top*
THC	*E70x14 Red Stripe Tires*
THE	*E70x70 Blackwall Tires*
TMC	*F70x14 Redline Nylon Tires*
SPS	*Special Solid Color Paint (Except Code A)*
SPR	*Special Black Code A Paint*
THD	*E70x14 White Stripe Tires*
TKM	*195Rx14 Whitewall Rayon Radial Tires*
TME	*F70x14 Blackwall Nylon Tires*

1969 Firebird Factory Installed Options

Sales Code	Option
A01	*Tinted Glass, Windshield only*
A02	*Tinted Glass, All*
A31	**Power Windows**
A46	**Power Bucket Seat**
A67	*Folding Rear Seat Back*
A90	Remote-Control Deck Lid Release
AS1	*Front Shoulder Belts (Convertible only after 01/01/69)*
B32	*Front Floor Mats*
B33	*Rear Floor Mats*
B93	Door Edge Guards
C06	*Power Top*
C50	**Rear Window Defogger**
C57	Power Flow Ventilation N/A with C60
C60	**Custom Air Conditioning**
D33	*Remote Control OSRV mirror*
D55	*Console N/A with Bench Seat*
DH5	Vanity Mirror L.H.
G08	*Vinyl Top*
G80	*Safe-T-Track Rear Axle*
G94-G83	*Performance Axle*
G95-G97	*Economy Rear Axle*
J41	*Custom Pedal Trim*
J50	*Power Brakes*
JL2	*Disc Brakes*
K02	*Heavy-Duty Fan*
K30	**Cruise Control**
K45	*Heavy-Duty Air Cleaner*
K82	*Heavy-Duty Alternator (C.S.I.)*
K96	55-amp Alternator
KB2	*Heavy-Duty Flex Fan*
L30	*350 Sport Group*
L67	*400-ci Ram Air IV*
L74	*400-ci Ram Air III*
L76	*Firebird HO Group*
M09	*Custom Gear Shift Knob*
M12	*Floor-shift for standard 3-speed (6-cyl only)*
M13	*Heavy-Duty 3-speed On Floor Transmission*
M20	*4-speed Transmission*
M21	*Close-Ratio 4-speed Manual Transmission*
M31	*Two-speed Automatic Transmission*
M38	*T.H. 350 Automatic Transmission*
M40	*T.H. 400 Automatic Transmission*
N10	*Dual Exhaust*
N30	*Deluxe Steering Wheel*
N33	*Tilt Steering Column*
N34	*Sport Steering Wheel*
N41	*Variable-Ratio Power Steering*
N65	*Space Save Spare Tire*
N95	*Wire Wheel Covers*
N98	*Rally II Wheels*
P01	*Deluxe Wheel Covers*
P02	*Custom Wheel Covers*
P17	*Spare Tire Cover*
PL3	*E78x14 White wall tires*
PX5	*F78x14 Blackwall Tires*
PX6	*F78x14 Whitewall Tires*
PY4	*F70x14 Whitewall Tires*
PY5	*F70x14 Redwall Tires*
RTT	*Standard Two-Tone Paint*
SPO	*Special Color*
STD	*250-ci 1-bbl*
STD.	*3-speed manual on column*
STT	*Special Two/Tone Paint*
U80	Rear Seat Speaker
U05	Dual Horns
U15	*Safeguard Speedometer*
U25	**Trunk Lamp**
U26	**Under Hood Lamp**
U30	*Rally Gauge Cluster with Tachometer*
U35	*Clock (N/A Available with Tachometer)*
U57	**Stereo Tape Player**
U58	**AM/FM Stereo**
U63	**AM Radio**
U69	**AM/FM Radio**
UA1	*Heavy-Duty Battery*
UB5	**Hood Mounted Tachometer**
V01	*Heavy-Duty Air Cleaner*
V64	**Instant-Aire Pump N/A with L67**
W55	Turnpike Group
W63	*Rally Cluster with Clock*
W66	*Firebird 400 Sport Group*
WS1	*Custom Front Seat Belts*
WS2	*Custom Front and Rear Shoulder/Seat Belts*
WS3	*Sprint Sport Group*
WS4	*Custom Trim Group*
WS4	*Trans AM Sport Group*
WS6	*Power Assist Group*
WS8	*Rally Group*
Y86	*Décor Group*
Y88	*Basic Group*
Y96	*Ride and Handling Package*

RPO A31/551/561 Power Windows

Optional all three years and was available on either the hardtop or convertible body styles. It consisted of a button switch that replaced the regulator handle. The switch was listed as part number 4432790 on each door. A special motor / regulator assembly was required to operated the door glass.

Power Window Regulator And Motor

Model Year	Driver's	Passenger's
1967	9709129	9709128
1968-1969	9719129	9719128

Power operated windows were also used on the rear quarter windows. These too required special regulators listed as part numbers 9709640 right-hand and 9709641 left-hand. These regulators were used all three years and with both the convertible and hardtop models. The switch was listed as part number 4432790 all three years and required a reinforcement part number 4722347.

RPO A46/564 Power Seats

With this option seat adjustment (up , down, forward and back) was accomplished by the touch of a switch, that used a button listed as part number 4309920. That controlled a small motor part number 4939203 was used to control the driver's side bucket seat. A wiring harness part number 7725119 was used to connect the motor and switch together. Power from the motor was transferred to a transmission assembly listed as part number 9705313. The motor and transmission assembly was used in all 1966-1969 Pontiac models with 4-way power bucket seat.

Special adjustment rails were required when the power seats were ordered. Only the driver's side was changed, the passenger's side bucket seat was not available as a power option, nor is there evidence to show that the option was offered on a car with a bench seat.

Power Seat Adjustment Rails

Model Year	Inner	Outer
1968	7793545	7793544
1969	8799147	8799146

RPO C50/404 Rear Window Defogger

Optional only on the hardtop coupe the rear window defogger blower motor part number3003268 for 1967 and was used in all Pontiacs except convertibles or wagons. In 1968 the part number was changed to 3014891, this continued to be used through 1969 but was used only on the Firebird models in 1969. The blower case should be painted 30-degrees gloss black. A grille (part number 3899746) assembly was used to cover the openings that had to be made in the package tray. A trim cover and spacers were used between the grille and the tray.

The switch assembly was mounted under the instrument panel and controlled with a round shaped knob part number 3899716 in 1967 and 1968 and was mounted in the instrument panel for 1969. The 1969 switch and wire leads were listed as part number 9797083.

RPO C60/582 Air Conditioning

A extensive option that consisted of many different components, and live coherently with the heater system. The blower motor requires a different unit than that when only heat is used. Part number 4914916 was used all three years. The evaporator unit was attached to the right-hand side of the firewall. Two different units were used. In 1967 part number 3013323 was used in 1968 the part number was changed to 3016308, this was used again in 1969 models except the Trans Am's. Trans Am's with air conditioning required a special evaporator listed as part number 302652, which was not used on any other model.

The expansion valve part number 5910619 was used all three years, as was the air inlet diaphragm. The throttle suction valve part number 5910704 was the same unit used on all 1966-1969 Pontiacs with air conditioning, with one exception the 1969 Trans Am which used part number 5910735, it was used in no other Pontiac application.

The condenser was listed as part number 3899793 for 1967 and 1968 and part number 3020098 for all 1969 models including Trans Am. Receiver assembly part number 3155342 was used in 1967 models while the 1968 and 1969 models used part number 3012944.

The air conditioning controls were located in the location of that for the heater and were integrated into each other. The control for 1967 models was listed as part number 3891797, 70303112 for 1968 and 7307832 for 1969.

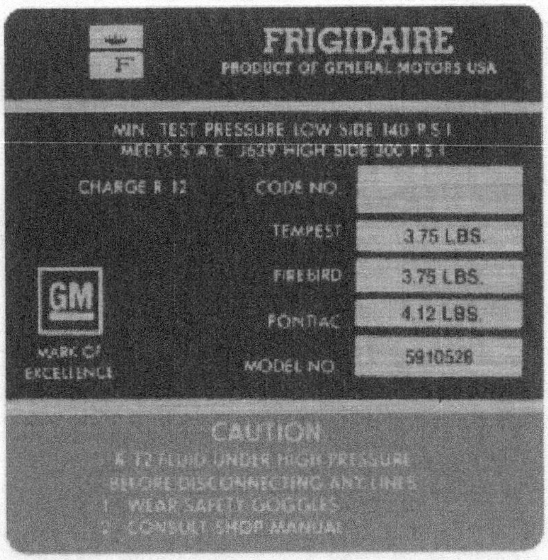

1967 compressor decal was red and black. Decal courtesy of Jim Osborn Reproductions

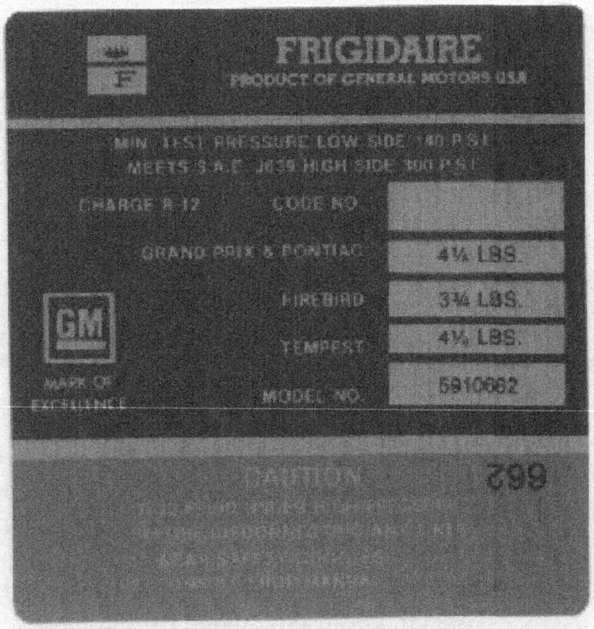

1968-1969 Compressor decal. Courtesy of Jim Osborn Reproductions

1967-1969 Hood Mounted Tachometers

Model Year	Engine Type	Part Number	Notes
1967-1968	Six Cylinder	6468670	Early 1968 6500rp, Redline
1967	V-8	6468675	
1968	Six Cylinder	6468956	Late 1968 5500 RPM Redline
1968	V-8	5657097	
1969	Six Cylinder	6469689	
1969	V-8	5657097	All V-8 Pontiacs

Hood mounted tachometer requires a hole to be cut into the hood to mount it.

RPO K30/441 Cruise Control

Available either from the factory or as a dealer-installed option ,this option found more use on full-size models, instead of the sporty Firebird. There were a few mandatory options with this package it required a V-8 power plant and automatic transmission. Because shifting gears is required as steady speed and steady resuming of speed could not be achieved and a six-cylinder did not supply enough vacuum to supply the package.

The speed regulator was placed under the hood and was listed as part number 9787978 for 1967 and 1968 and was used all 1967-1968 Pontiac models that were equipped with cruise control. The part number was changed to 9797504 in 1969; this part was used on all 1969-1971 Pontiac models. A bracket part number 9777850, held the regulator and was used all three-years. The lower speedometer cable was routed from the transmission to the regulator; the upper cable was routed from the regulator to the speedometer head.

A 5/16-in. diameter rubber hose was routed from the regulator to a small plastic "T" connector part number 3887105. From the T the hose was split and a 5/16 in. diameter hose was routed to the servo assembly part number 9787977, and a ¼ in hose was routed from the T-connector to release switch part number 9794682.

A slightly smaller ¼-in. diameter hose was routed from the regulator to the engine vacuum source. A wiring harness was routed from the regulator to the dash. Other under hood components included special brackets, bead chain to hold the carburetor throttle in position. A switch on a special turn signal stalk regulated speed. Part number 6465211 was used on the 1967 Firebird only, while part number 6465256 which was used on 1968, and 1969 models, was used on all 1969-71 Pontiac models with cruise control.

RPO U25/421/671 Under Hood Lamp
RPO U26/401/652 Trunk Lamp

Available as part of the Light Package or as individual options, it is correct for your Firebird to have only trunk compartment light and not an underhood lamp, provided the lamp package was not ordered. The lamp package varied slightly each year but included the under hood lamp, trunk compartment lamp, ignition lamp, and courtesy lamps in convertible models.

The under hood lamp and the trunk lamp were also available as dealer installed options but only on the 1969 models.

Lamp Part Numbers

1967	
Under Hood	3878110
Trunk Lamp	3878110
1968	
Under Hood	3916764
Trunk Lamp	3878110
1969	
Under Hood	9796394
Trunk	5405081 *
	3878110#

*-Factory Installed
#- Dealer Installed

RPO V64 Vacuum Tire Pump

A unusual and rare option was available only on 1969 models, it used the vacuum air that the engine created to run a small air pump part number 7804055 and was optional on all powerplants except the Ram Air models. An adapter part number 7805030, was used on all powerplants and held in place with a bracket part number 9794236 for six-cylinders and 9794235 for V-8 models. A 15-feet long hose, part number 7805015 allowed you to reach each tire on the car. A 11-1/2 in long ½ I.D. hose ran from the adapter to the vacuum fitting.

Vacuum Fittings

250-ci 1-bbl	9794237
250-ci 4-bbl	9794238
350-ci or 400-ci Except Ram Air	9794239

Appendix 1 Jacking Decals

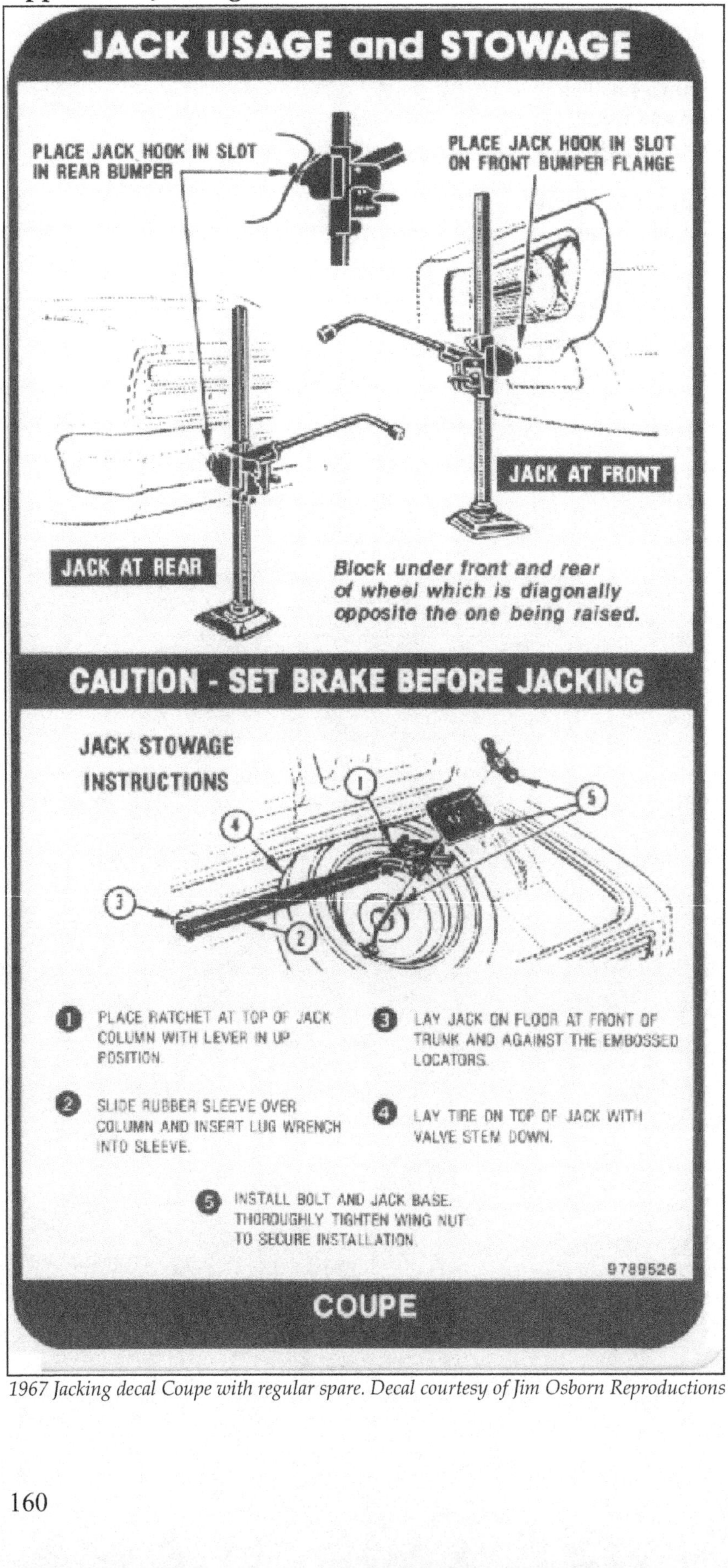

1967 Jacking decal Coupe with regular spare. Decal courtesy of Jim Osborn Reproductions

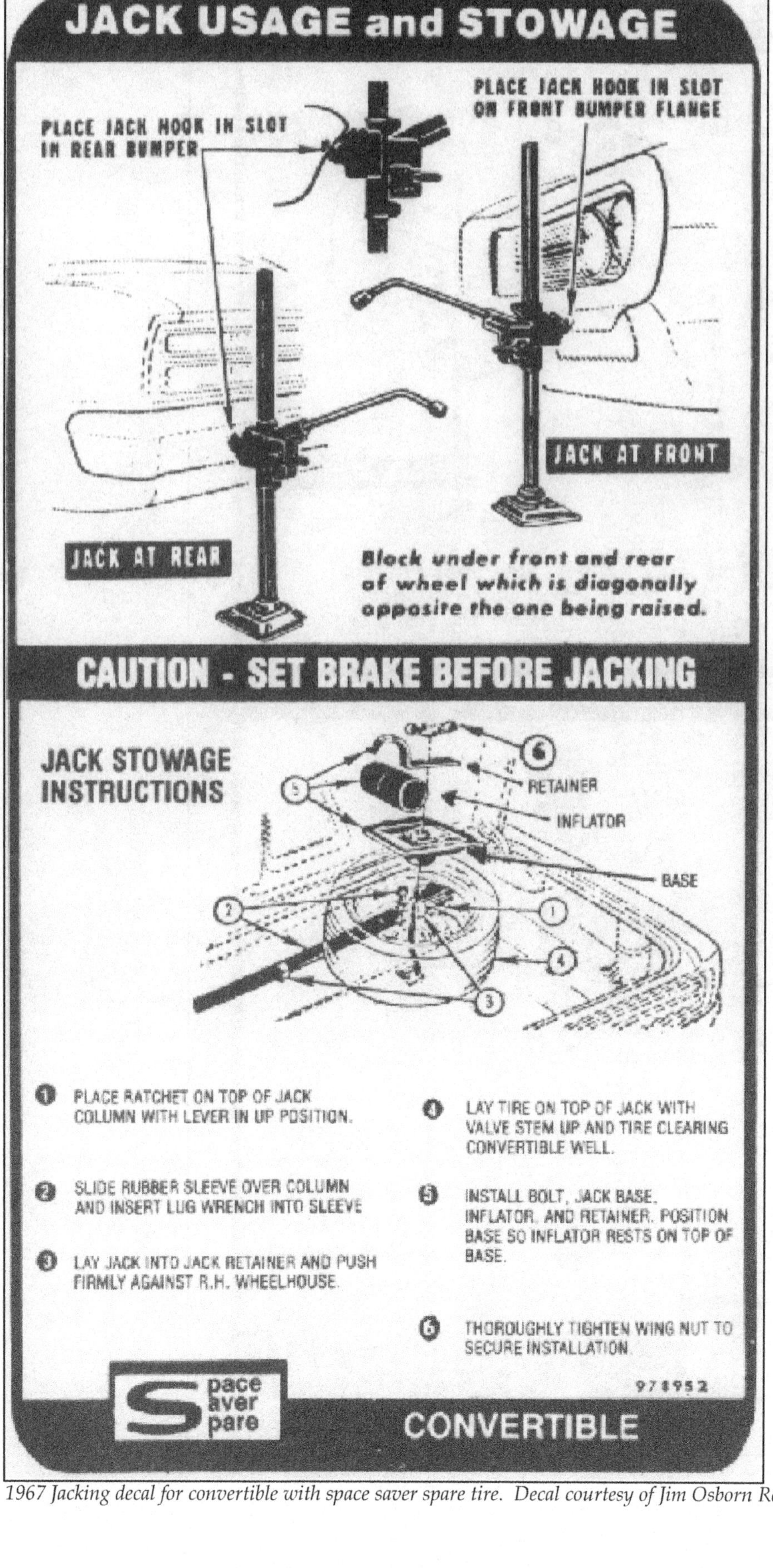

1967 Jacking decal for convertible with space saver spare tire. Decal courtesy of Jim Osborn Reproductions.

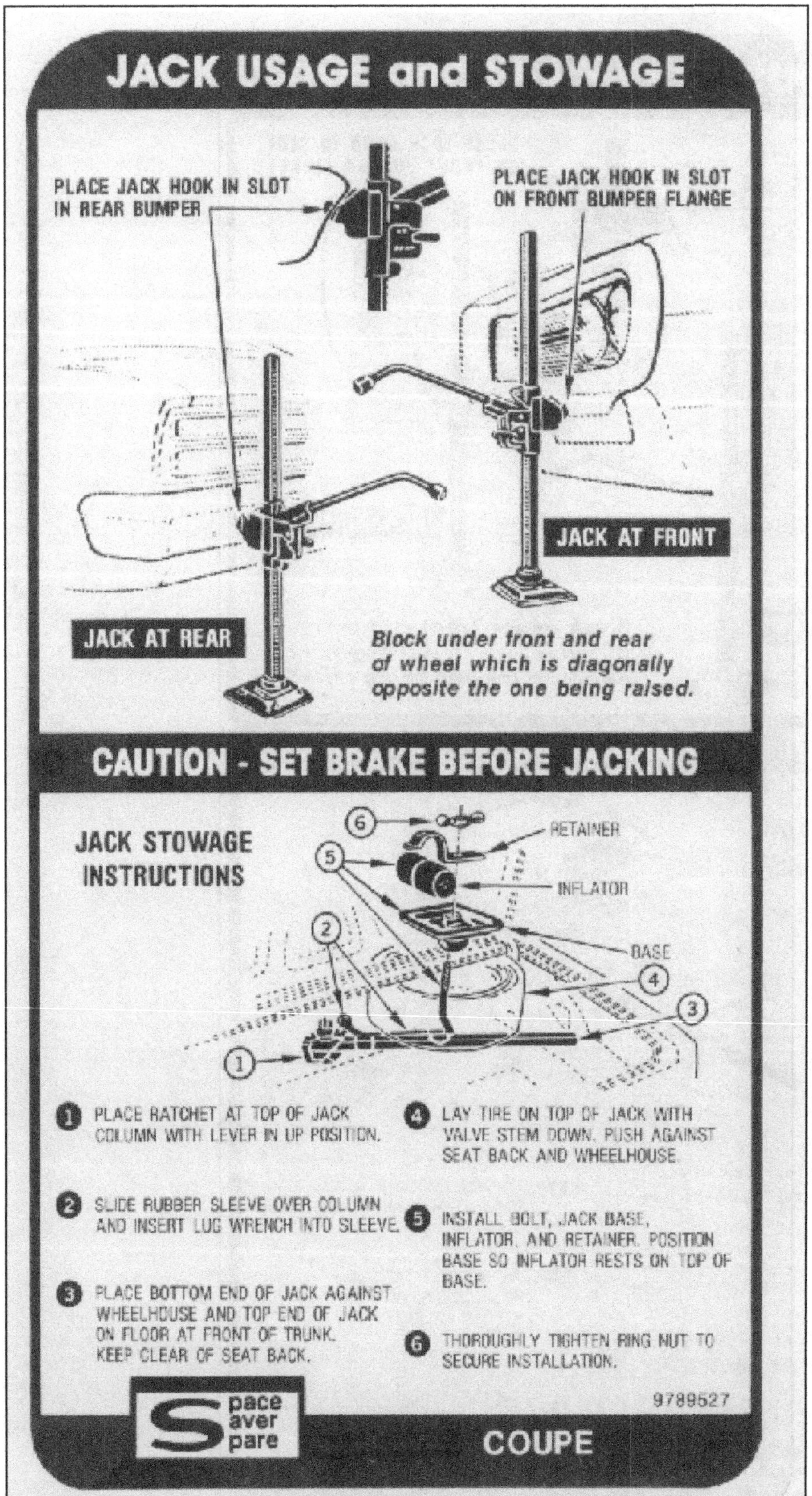

1967 Jacking decal for Coupe with space saver spare tire. Decal courtesy of Jim Osborn Reproductions.

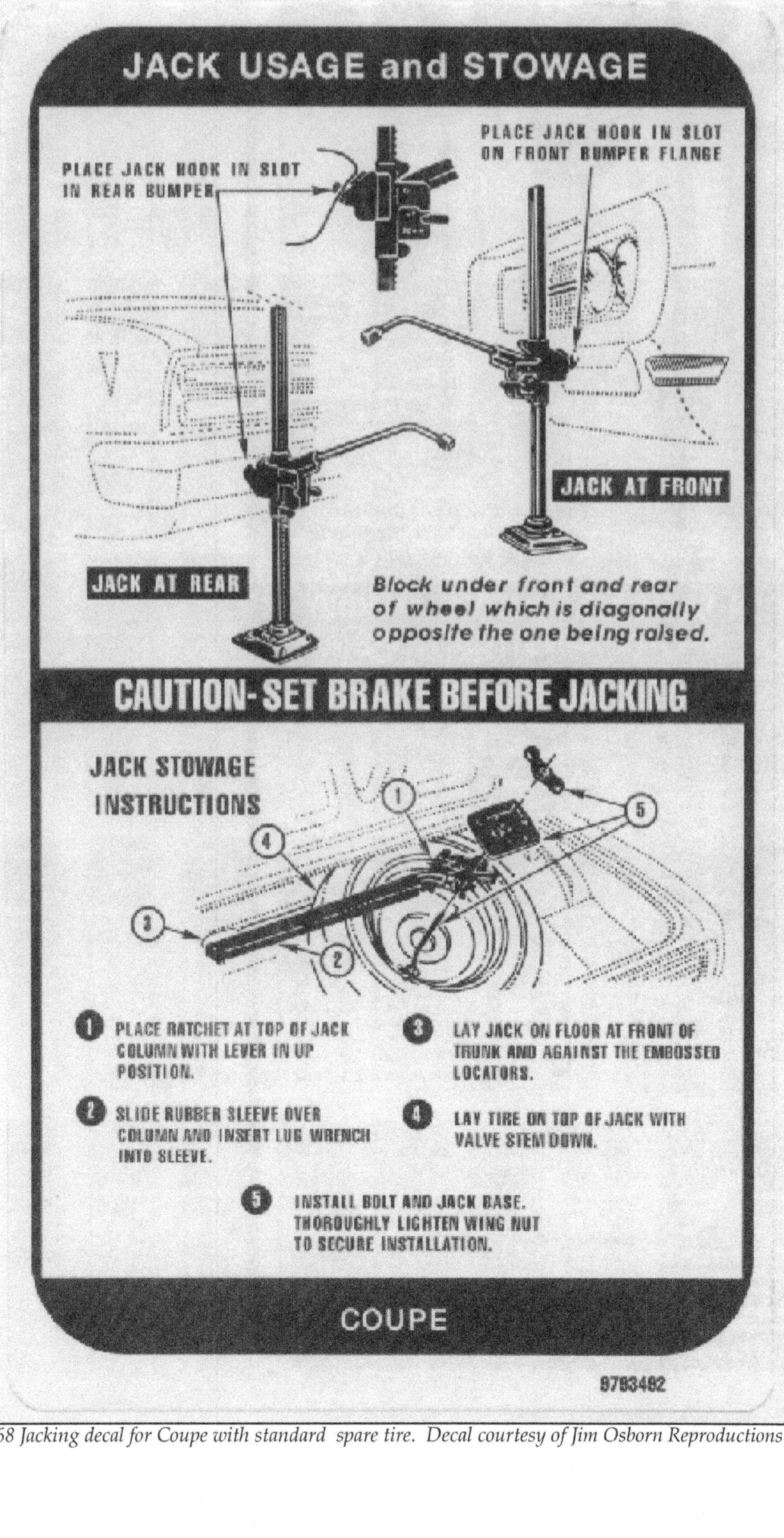

1968 Jacking decal for Coupe with standard spare tire. Decal courtesy of Jim Osborn Reproductions.

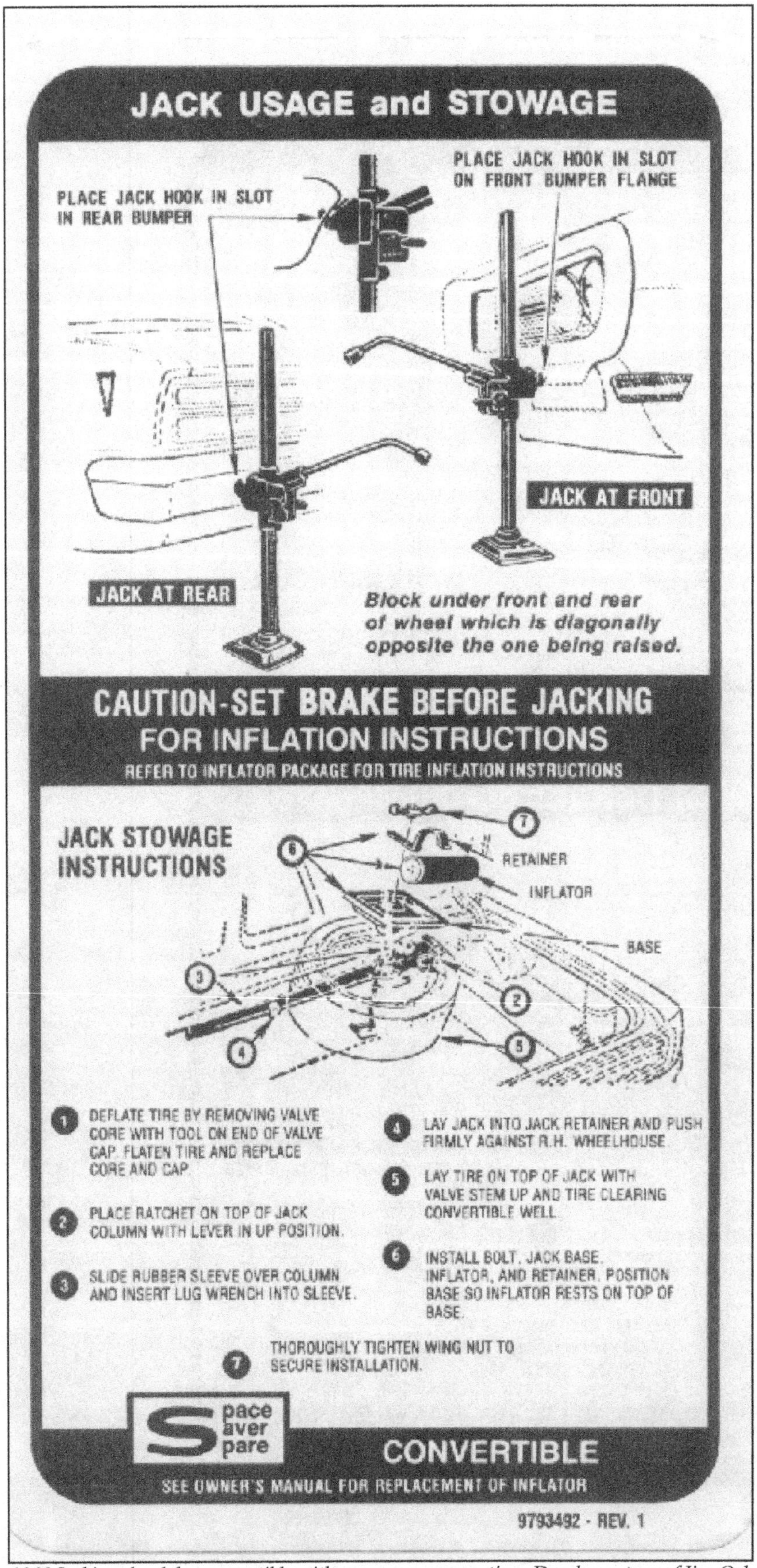

1968 Jacking decal for convertible with space saver spare tire. Decal courtesy of Jim Osborn Reproductions.

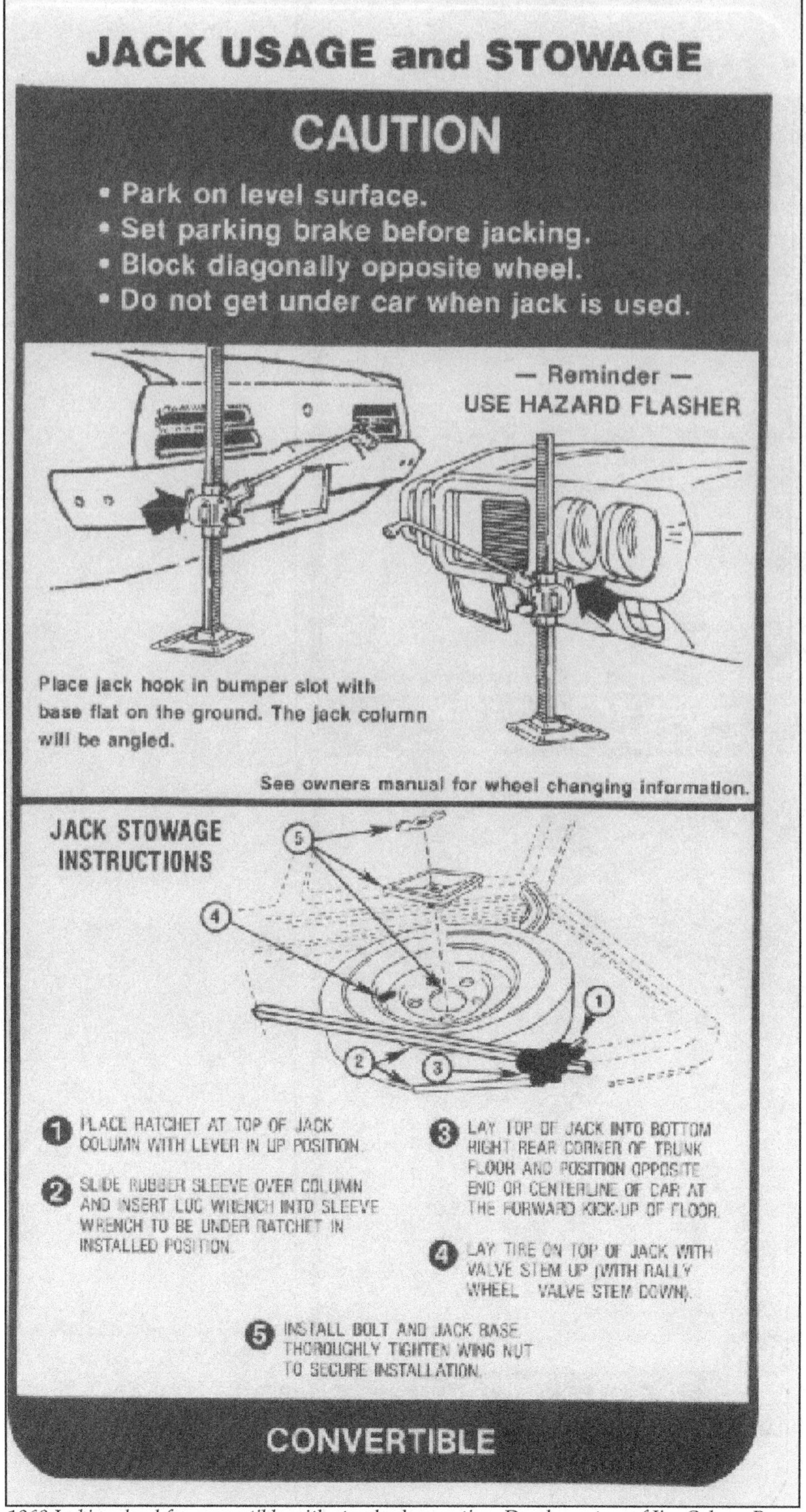

1969 Jacking decal for convertible with standard spare tire. Decal courtesy of Jim Osborn Reproductions

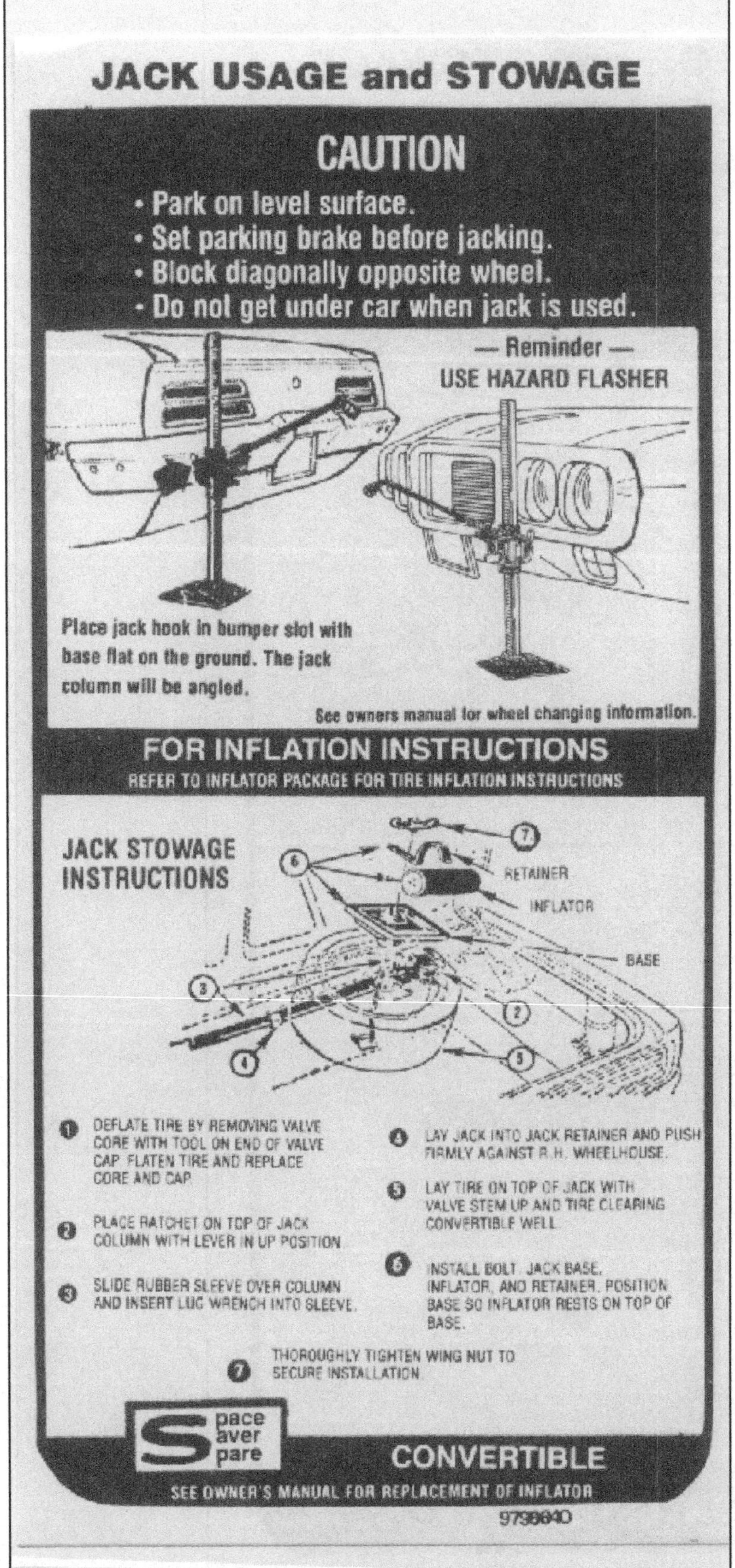

1969 Jacking decal for convertible with Space Saver spare tire. Decal courtesy of Jim Osborn Reproductions

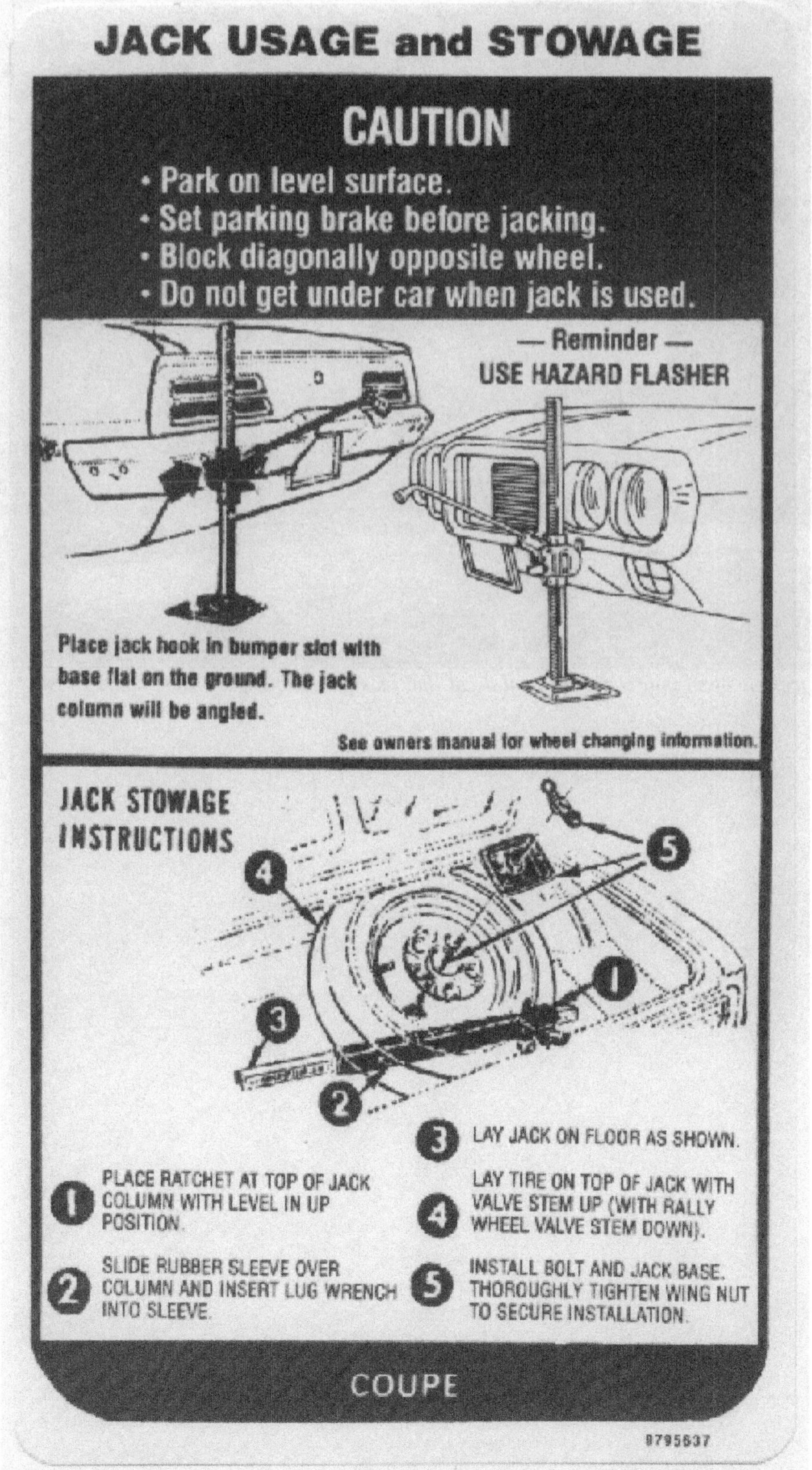

1969 Jacking decal for coupe with Standard spare tire. Decal courtesy of Jim Osborn Reproductions

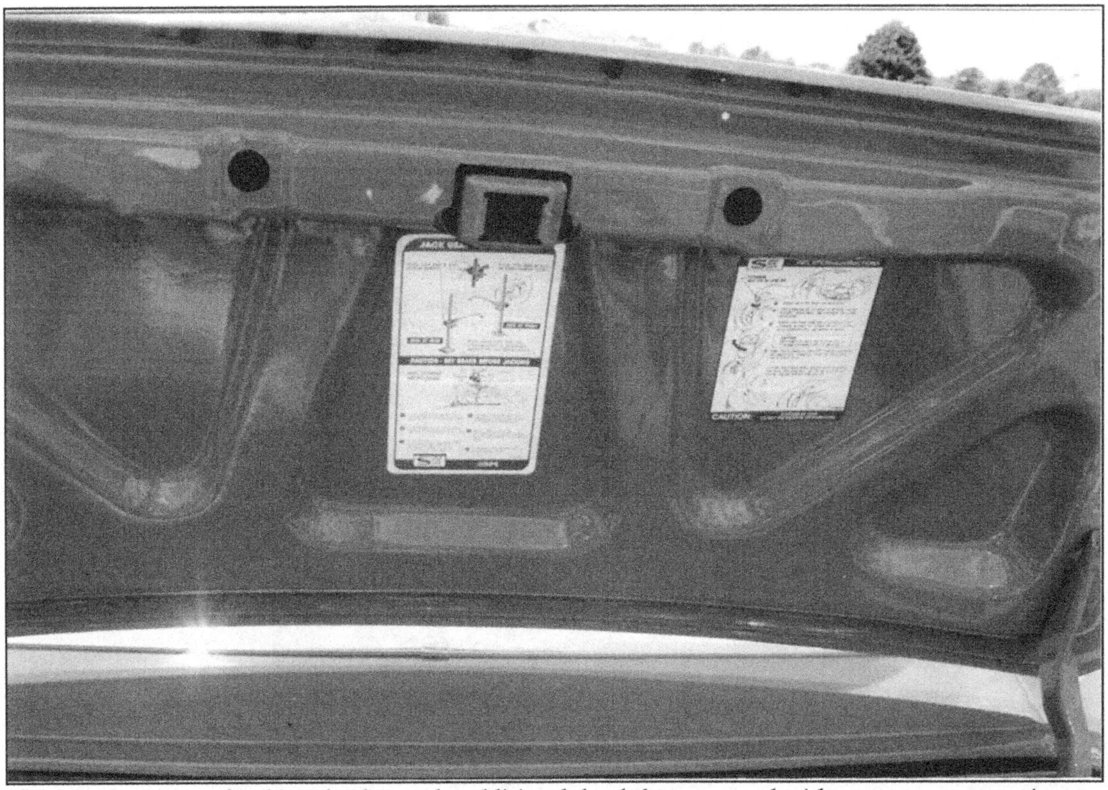

TYpical placement of Jacking decal note the additional decal that was used with space saver spare tires.

Appendix 2: Production Totals

Model Year	Production		Notes
1967	82,560		
1968	107,112		
1969	87,011		Includes Trans Am

Total 1967-69 Firebird Production **276,683**

Production Total By Body Style

Model Year	Hardtop	Convertible
1967	67,032	15,528
1968	90,152	16,960
1969	75,362*	11,649*
Total	**231,857**	**44,129**

*Includes Trans Am

Production Total By Model

Model Year	Standard	Deluxe
1967	34,380	48,180
1968	71,851	35,261
1969	87,011*	Series deleted
Total	**193,242**	**83,441**

*Includes Trans Am

Production Total Powerplant/Transmission

Model Year	Six Cylinder		V-8	
	Manual	Automatic	Manual	Automatic
1967	8,221	9,443	19,750	45,146
1968	8,744	9,750	24,166	64,452
1969	9,820	11,073	25,088	49,585
Total	**26,785**	**30,266**	**69,004***	**159,183***

*Includes Trans Am

Production Total By Engine Size

Model Year	Six Cylinder	326/350V-8	400-V-8
1967	17,664	46,196	18,700
1968	18,494	67,282	19,237
1969	20,893	52,930	13,188*
Total	**57,051**	**166,400**	**51,117**

*Includes Trans Am 326- 1967 350-ci 1968-69

Ram Air Production Total

Model Year	Ram Air	Ram Air II	Ram Air II	Ram Air IV
1967	65	0	0	0
1968	413	110	0	0
1969	0	0	1,509	157
Total	**478**	**110**	**1,509**	**157**

Total Ram Air Cars- **2,254**

Trans AM Production Total By Body Style

Model Year	Hardtop	Convertible
1967	0	0
1968	0	0
1969	689	8
Total	**689**	**8**

Total Trans Am 697

Trans Am Production Total by Power plant/Transmission

Ram Air III Manual	Ram Air III Automatic	Ram Air IV Manual	Ram Air IV Automatic	
520	114	46	9	No convertible Ram IV's were built.

Top Ten Rarest 1967-1969 Firebirds

Model	Total	Percentage of Production
1. 1969 Trans Am Convertible	8	-.01
2. 1967 400 with Ram Air	65	-.01
3. 1969 Trans Am with Ram Air IV	55	-.01
4. 1968 Firebird with Ram Air III	110	-.01
5. 1968 Firebird with Ram Air *	413	-.01
6. 1969 Trans Am hardtop	689	-.01
7. 1968 Firebird Deluxe six cyld. Manual transmission	1,216	.01
8.1968 Firebird Deluxe with six cyld. Automatic transmission	1,309	.01
9.1969 Firebird with Ram Air IV	1,509	.01
10. 1969 Firebird Convertible	11,641	13

Appendix 3 Firebird Part Suppliers

Name/Address	Phone/ Website	Supplies
Jim Osborne Reproductions **101 Ridgecrest Dr.** **Lawrenceville, GA 30045**	Phone:770-962-7556 www.osborn-reproductions.com	Quality reproduction decals
TexasTransAms.com 4588A Kennedale New Hope Road Fort Worth TX 76140	Phone: 817-563-2121 www. texastransams.com	A selection of Firebird parts and restorable cars for sale. Also a Firebird salvage yard
Warpath Restoration Parts Parma OH 44130	Phone 440-845-4612 www. warpath-parts.com	
The Parts Place 217 Paul St Elburn, IL 60119	630-365-1800 www.thepartsplaceinc.com	A section of Firebird parts
Classic Exhaust 805 Program Parkway Geneva, OH 44041	Phone: 440-466-5460	Exhaust parts
Pontiac Parts	Phone: 800-447-2886 www.pontiacpart.com	
Franks Pontiac Parts Ramona, CA 92065	Phone: 760-789-0669 www.frankspontiacparts.com	Used and new parts
Classic Industries 18460 Gothard Huntington Beach, CA 92648	Phone 800-854-1280 www.classicindustries.com	Selection of reproduction parts.
National Parts Depot	Phone 352-378-9000 www. npdlink.com	Great selection of reproduction parts
Faxon Auto 3901 Carter Ave. Riverside, CA 92501	Phone: 1-800-458-2734	Large section of automotive literature
Paper N Parts 5299 US Hwy. 127 Bryan, OH	Phone: 419-636-8674	
Original Parts 5252 Bolsa Ave. Huntington	1-800-243-8355 www.gmrestoparts.com	Larges selection of reproduction parts
Pontiac Oakland Club Internat. P.O. box 14057 Bradenton, Fl 34280	941-792-4992	Pontiac club
Year One P.O. Box 521 Braselton, GA 30517	1-800-932-7663 www.yearone.com	Selection of reproduction parts.
CST Salvage **Aurora MO**	417-678-7305 417-678-6994	Salvaged Firebirds in Yard
Henderson Auto Salvage **Monett MO**	417-235-3719	Salvaged Firebirds in Yard

W

Made in the USA
Monee, IL
05 November 2026

67985672R10096